Perovskite based Materials for Energy Storage Devices

Edited by

Inamuddin[1], Maha Khan[1], Mohammad Abu Jafar Mazumder[2,3]

[1]Department of Applied Chemistry, Zakir Husain College of Engineering and Technology, Faculty of Engineering and Technology, Aligarh Muslim University, Aligarh-202002, India

[2]Chemistry Department, King Fahd University of Petroleum & Minerals, Dhahran 31261, Saudi Arabia

[3]Interdisciplinary Research Center for Advanced Materials, King Fahd University of Petroleum & Minerals, Dhahran 31261, Saudi Arabia

Published by **Materials Research Forum LLC**
Millersville, PA 17551, USA

Published as part of the book series
Materials Research Foundations
Volume 151 (2023)
ISSN 2471-8890 (Print)
ISSN 2471-8904 (Online)

Print ISBN 978-1-64490-272-1
eBook ISBN 978-1-64490-273-8

Distributed worldwide by

Materials Research Forum LLC
105 Springdale Lane
Millersville, PA 17551
USA
https://www.mrforum.com

Manufactured in the United States of America
10 9 8 7 6 5 4 3 2 1

Table of Contents

Preface

Organic-Inorganic Perovskite Based Solar Cells
M. Rizwan, A. Ayub, S. Urossha, M.A. Salam, M.W. Yasin, A. Manzoor,
S. Mumtaz 1

Organometallic Halides-Based Perovskite Solar Cells
Uzma Hira, Muhammad Husnain 33

Perovskite Based Ferroelectric Materials for Energy Storage Devices
M. Rizwan, A. Ayub, T. Fatama, H. Hameed, Q. Ali, K. Aslam, T. Hashmi 67

Techniques for Recycling and Recovery of Perovskites Solar Cells
Somi Joshi, Kanchan Chaudhary, Kalpana Lodhi, Manjeet Singh Goyat,
Tejendra K. Gupta 89

Lead-Free Perovskite Solar Cells
Mridula Guin, Riya Singh 111

Technical Potential Evaluation of Inorganic Tin Perovskite Solar Cells
Lutfu Sagban Sua, Figen Balo 155

Keyword Index
About the Editors

Preface

In the third generation of photovoltaic systems, the invention of organic/inorganic-based hybrid perovskite materials has transformed the technology. The efficiency of perovskite materials for solar cells has increased almost by 22% in the last decade. There are several reasons for the power conversion efficiency (PCE's) strong optical absorption, balanced charge transfers, and thicker diffusion layer. The n-i-p and p-i-n with mesoporous or planar heterojunction perovskite solar cell architectures have emerged as the two most common types. Perovskite-based supercapacitors now have some of the desirable light harvesting features, such as higher absorption coefficients, longer carrier lifespans and diffusion lengths, bipolar transport of carrier, and superficial defect levels, thanks to the unending efforts to enhance perovskite solar cells. There have also been significant improvements in the cell architecture, working procedure, layer development, and bandgap correction of perovskite solar cells. This book overviews the reason for perovskite supercapacitors to be a commercially viable future power solution in the next few years due to their increased efficiency, cost-effective raw materials and processing, superior optoelectronic characteristics, and simple device construction procedures. The summaries of the chapters are given below:

Chapter 1 discusses the organic-inorganic perovskite solar cells (PSCs), their structure, and optoelectronic properties. The influence of individual and mixed substitution at A, B, and X sites on properties such as band gap and performance of these solar cells is also discussed. Various materials used for the electron transport layer, hole transport layer, and absorbing layer and their fabrication methods such as spin coating are discussed. Challenges and future perspectives are also discussed.

Chapter 2 throws light on modern trends in solar cells (SCs). The main theme of this content is on the organometallic halides-based perovskite solar cells (OMHP-SCs) and their structure, parameters, classification, and photovoltaic effect. Environmental instability, power conversion efficiency (PCE), and advancement through passivation techniques in OMHP-SCs are also discussed.

Chapter 3 discusses ferroelectric-based perovskites that are mainly used in storage devices such as capacitors, and fuel cells. Lead-based perovskites such as niobate and lanthanum-based ferroelectric perovskites are discussed. Lead-free perovskites such as barium titanate-based, bismuth-based, and alkaline niobate-based are also highlighted. Different energy storage devices and ways to improve their efficiency with methods like chemical substitution are discussed.

Chapter 4 discusses the roadmap and need for recycling perovskite-based solar cell modules, the toxicity behavior of lead present in the solar cells, and the recycling of different parts of perovskite solar cells. It also sheds light on the cost analysis of recycling and future challenges.

Chapter 5 presents the current status of lead-free perovskite solar cells (PSCs) and their prospects. Various strategies to enhance photovoltaic efficiency are discussed in detail. Several materials are highlighted that can be explored further to progress in this area. Lastly, different fabrication processes of high-quality PSC film and associated challenges are discussed.

Chapter 6 analyses the most effective solar cell among the latest inorganic tin perovskite solar cells by AHP methodology. The inorganic tin perovskite photovoltaic cells as the selection problem with multiple-criteria decision-making methodology have been first examined by this research.

Perovskite based Materials for Energy Storage Devices
Materials Research Foundations 151 (2023) 1-32

Materials Research Forum LLC
https://doi.org/10.21741/9781644902738-1

Chapter 1

Organic-Inorganic Perovskite Based Solar Cells

M. Rizwan[1*], A. Ayub[2], S. Urossha[1], M.A. Salam[1], M.W. Yasin[1], A. Manzoor[1], S. Mumtaz[1]

[1]School of Physical Sciences, University of the Punjab, Lahore, Pakistan

[2]Department of Physics, University of the Punjab, Lahore, Pakistan

*rizwan.sps@pu.edu.pk

Abstract

Traditional silicon-based solar cells have dominated the photovoltaic industry for quite some time now. Alternatives of these solar cells are being researched such as hybrid organic-inorganic perovskite-based solar cells, which are cost-effective and have the potential to achieve higher efficiency and performance. The characteristics of these perovskites can be controlled via substitution at A, B, or X sites since all these have a great impact on the overall performance. The components of perovskite solar cells (PSC) along with materials used for these layers, and fabrication techniques that give the optimized efficiency, challenges, and future perspective faced by perovskite solar cells are also deliberated.

Keywords

Lectron Transport Layer, Spin Coating, Organometallic Materials, One-Step Deposition Perovskites, Hole Transport Materials

Contents

Organic-Inorganic Perovskite Based Solar Cells..1

1.   Introduction..3

2.   Silicon Solar Cells (SSCs)..5

3.   Perovskites-Based Solar Cells (PSCs)..6

      3.1   Structure of PSCs...7

      3.2   Optoelectronic Properties Of PSCs 9

      3.3   Influence of A, B, and X site 10

3.3.1 A-Site 10

3.3.2 B-Site 10

3.3.3 X-Site 11

4.   Mixed Concentration of Perovskite Absorbing Layer **11**

4.1   A-site 12

4.4   Mixed B-Sites Cations 13

4.5   X-Site 13

5.   Requirements for Each Layer **14**

5.1   Electron Transport Layer 14

5.1.1 Different ETL Material Used In Perovskite Cells 15

5.2   Hole Transporting Layer 15

5.2.1 Hole Transporting Material (HTM) 16

5.2.2 Inorganic P-type semiconductors as HTMs 16

5.2.3 Organometallic HTMs 17

5.3   Absorbing Layer 17

5.3.1 Preparation Method of The Perovskite Light Absorbing
       Layer 17

6.   Fabrication Techniques **18**

6.1   One-Step Deposition 18

6.2   Two-Step Deposition 18

6.3   Vapor Deposition Method 20

6.4   Spin Coating 21

6.4.1 One-Step Spin Coating 21

6.4.2 Two-Step Spin Coating 21

6.5   Thermal Vapor Deposition 21

7.   Challenges in Perovskite-Based Solar Cells **22**

7.1   Stability Challenges 22

7.2   Thermal Effect 23

7.3   Toxicity 23

7.4   J-V Hysteresis 24

8.   Efficiency of Perovskite **24**

9.   **Future Perspectives** **26**

Conclusion...**26**

References **27**

1.   Introduction

The energy crisis has caused dire consequences for the world economy. With time, energy needs have become the biggest problem because of the increase in demands day by day. The cheapest way of producing electricity is via water but due to failed economies, lack of water in dams, and poor management of rainwater, we are facing an extreme energy crisis. The thermal power electricity, produced by oil and gas is highly expensive. Nuclear power stations are another means of energy, but nuclear reactors are highly affluent to set up for developing countries. Consumption of electricity is far greater than the production of electricity. There is an indispensable necessity to adopt new conduct to overcome this energy crisis.

Solar cells are the primary devices to deal with energy crises. These are based on the principle of the photovoltaic effect, a phenomenon that directly converts incident light into electricity via a semiconductor device such as a p-n junction. An electric field is set up by combining the two types of semiconductors. In the region of the junction where the field is formed, the positive holes migrate to the negative n-side and negative electrons migrate to the positive p-side. Because of the field setup, the positively charged particles move in one direction and negatively charged particles in the other direction. Light is composed of minuscule packets of electromagnetic radiation or energy which are known as photons [1]. Whenever a suitable wavelength of light falls on the semiconducting materials, energy is transferred from the photon to an electron causing it to jump to a higher energy level known as the conduction band (CB), thus producing current.

Due to photovoltaic installations, 100GW (Giga-watts) of electricity was produced at the end of 2030. 85% are using crystalline silicon, mostly cadmium telluride/cadmium sulfide in being polycrystalline thin film solar cells. Thin film solar cells have short energy reproduction time although these cells are cheaper. They have some disadvantages, they may become crucial while working in the terawatt range. Mostly they are composed of rare elements e.g. tellurium (rare as gold), gallium, and indium [2].

A new arrival to the Photovoltaic field has extended the solar conversion efficiencies to 15%, these solar cells are based on perovskite-structured semiconductors like methylammonium lead iodide ($CH_3NH_3PbI_3$). These perovskites have high energy charge carrier mobilities because electrons and holes can travel huge distances when current is to be removed, as they don't give up their energy in the form of heat within the cell. The perovskite cells have great importance as they are set up by spin coating (low-temperature solution method). For the eventual manufacturing of these cells, the easiest way of deposition is important. The lowest temperature (below 100°C) solution refined films have smaller diffusion lengths, although larger diffusion lengths can explain the photocurrents obtained with materials and high quantum efficiencies.

Three considerations that will affect the performance of these perovskites solar cells are mainly:- first is energy conversion efficiency. This aspect is in good shape because, with an efficiency of 15.4%, it works for several years and is near the theoretical limit, which is surprising, second is cost, this consideration is more perplexing because it involves energy reproduction time and energy cost as well as the presence of raw material. The minimum energy requirements in cell fabrication are translated through the low-temperature solution methods [2]. There exist no rare elements involving gold that have cheaper contact material alternatives. While there are some other possible replacements like tin, it will not be a key issue even in commercial cells if Pb is required.

The third consideration is stability. The study shows around a 20% decrease in efficiency after 500 hours. In fact, it is highly encouraging, when we observe the newly made inorganic cell stability; we observe the cause of optimization is that they can be made commercially important over the years.

Perovskites material like methylammonium lead halides (MAPbI) and all inorganic Cs-Pb halides are simple to manufacture and cheap to produce. Perovskites solar cells have become commercially attractive with the potential of having very low production costs and higher efficiencies. Perovskites have become a swiftly advancing solar technology in the modern era. The recent work on perovskites has been influenced by the absorber material based on methylammonium lead halides [3].

Perovskite performance can be extended in the solar spectrum to respond to a variety of wavelengths by controlling the material composition. The absorbing power of perovskites can be improved by introducing modifications in their composition. Perovskites solar cells of certain compositions may convert visible and ultraviolet radiations into electrical energy. There is a possibility that two perovskites having different compositions are combined, and they produce a perovskite-only tandem. Perovskites-only tandem solar cells can be fabricated on adjustable substrates with high power-to-weight ratios, which could turn competitors in tragedy retort and in defense operational energy areas. Alternatives are also being explored to reduce, evaluate, and potentially eliminate toxicity, and alleviate environmental concerns.

2. Silicon Solar Cells (SSCs)

Our Earth's crust is made up of 27.7% silicon which is the 2nd most plentiful element in the crust after oxygen. Silicon is also known as silica. Pure silicon is the fundamental component of traditional solar cells. SSC technology is referred to as the 1st generation panel as it has gained ground already in the 1950s. Even more than 90% of the current solar cell technology is based on silica. Pure crystalline silicon is a semiconductor material thus; it is a poor conductor of electricity. To overcome this issue, other atoms are incorporated with silicon atoms in a solar cell to improve silicon's conducting ability. After the addition of other atoms, silicon becomes a pure conductor and captures sun energy properly which is converted into electrical energy.

When solar cells were utilized for energy production, they set a revolution of change in motion. Solar cell technology is a continuously changing and evolving field which changed the face of energy production. For natural warmth, in the very early days, sunrooms were built. In 1839, the photovoltaic cell was discovered by a physicist, Edmond Becquerel, when he was experimenting with a cell in a conducting solution and the cell was made of electrodes. Willoughby Smith, in 1873 suggested that selenium could also function as a photoconductor [4]. After three years, Richard Evans and William Grylls applied the photovoltaic principle which was discovered by Becquerel. After 50 years, the photovoltaic effect was discovered by an American, Charles Fritz, who introduced the 1st working selenium cell. However, in modern solar panels, we use silicon in cells. The 1st SSC was proclaimed in 1941 and had an energy efficiency of < 1%.

In 1980, Si-based solar cells were prepared from crystalline silicon wafers. The evolution of work on the ultra-thin silicon substrate was energized by the very expensive cost of raw material polysilicon. In 2008, the price of silicon wafers was considered more than photovoltaic modules at 30% which was about 60% of the price of wafers obtained from wafering and bloom growth. In fact, a major selection of energies was used to fabricate photovoltaic modules to increase the energy reproduction time by silicon wafer manufacturing. So, prices of silicon were reduced in 2008 from about 400 to 30$/kg with the continuous production of silicon solar cells in Korea and China in 2013. Thus, introducing an advanced wafer technology and expanding it into the present process chain remains an enormous challenge. Silicon solar panels are quite rigid and fragile which are not ideal properties for transportation. The fragments of silicon solar panels are much more expensive as compared to alternative options in solar technology techniques. Solar cells are made up of a material that is not abundant in nature and these elements are a necessity in this system. Silicon solar cells are very expensive. Silicon solar cells have less efficiency as compared to perovskites solar cells.

3.   Perovskites-Based Solar Cells (PSCs)

Perovskites are a family of materials named minerals that have specific crystal structures. These materials are used to make solar cells having high-performance potential at lower costs. Due to their high efficiency, they have shown remarkable progress in this modern era. In a short time, they have become highly efficient. The majority of the recent work on perovskites is based on absorber material based on methylammonium lead halide. Perovskites except III-V technologies have exceeded all thin film technologies in power conversion. By tuning the material composition, perovskites can be used to respond to different wavelengths. Perovskites solar cells can be used to convert visible and ultraviolet light into electricity very efficiently and thus are the best hybrid tandem contenders for absorbers like crystalline silicon that alter infrared light efficiently. Two different compositions of solar cells can be joined together to constitute a perovskite-only tandem [5]. These tandem photovoltaics (PV) may lead to applications having higher efficiency and cost-effectiveness. These PSCs are competitive in disaster response, in mobile and defense operational energy areas, and can also produce on flexible substrates with power-to-weight ratios.

Among different methods used to fabricate perovskite solar cells, two major methods are:

(i) Sheet to sheet: In this type, device layers are deposited on a rigid substrate; in the complete solar surface, they act as a front surface. It is commonly used in cadmium telluride modules.

(ii) Roll to roll: In this type, the layers are deposited on a flexible substrate. In the complete module, they can be used as an exterior as well as interior surface.

Researchers have tried to use this type of device but they failed for significant commercial traction. But they are used in chemical and photographic film and paper products like newspapers. Various device structures and fabrications are available for further progress and perovskites solar cells.

In 1839, calcium titanium oxide ($CaTiO_3$) also known as calcium titanate was first discovered by German mineralogist Gustav Rose. At the end of the 18th century, work began on solar cells. It was discovered that selenium can generate electricity when interacting with light. It was seen that selenium-based solar cells were not able to store electrical energy through a device considering sunlight as a source. Charles Fritts explained the solar cell firstly which was made up of selenium wafer in 1883. The selenium solar cells were gold coated and fabricated with a conversion efficiency between 1 and 2%. The first photovoltaic cell constructed on silicon was invented by Bell laboratories in 1954. Gerald Pearson, Calvin Fuller, and Daryl Chapin explained that fabricated silicon photovoltaic cells can store power as electrical energy through sunlight. This energy was sufficient to run an electrical instrument. This silicon solar cell had a modified 11% efficiency rather than 4%. In 1958, the Si-solar cell was considered in satellites for space purposes as an energy source. After the application of PV in satellites, it was used in off-grid power stations. In the 1970s, Elliot Berman prepared an economic PV. This application in grid stations was highly expensive. These PSCs have applications in domestic as well as railroad crossing. After that, the researchers started to work on thin-film solar cells. In 1980, the 1st thin film solar cell was fabricated at the University of Delaware by using cadmium sulfide and copper sulfide, the efficiency of which was more than 10%. In 2009, the hybrid organic-inorganic lead perovskite was first utilized for PV solar cells with visible light, having an efficiency of 3.8%. The first perovskite based on a stable solid-state was proclaimed with an efficiency of 9.7% in 2015, the research on perovskite solar cells gained momentum and their efficiency reached 21.5%, in the PV industry perovskites are considered remarkable candidates within a short time [5]. Perovskites solar cells have a greater tendency to absorb at higher wavelengths. PCE (power conversion efficiency) of perovskites is high compared to traditional solar cells and reaches up to 25.5% efficiency.

PSCs are mainly composed of PV absorbers, counter electrodes, and transport layers. In perovskites solar cells, the efficiency is determined by energy loss within the bulk. The manufacturing of PSCs absorbers doesn't need complex equipment. These absorbers as compared to silicon can be transferred at a lower temperature. They are not highly expensive as compared to silicon-based solar cells, where Si wafers are essential components of solar cells which make up 70% of the instrument. In DSSC (dye-sensitized solar cells), the PSCs were integrated first. The conventional dyes were replaced by using these solar cells. As compared to ruthenium, the PSSC (perovskites-sensitized solar cells) have a greater ability to produce electrical energy through photons. Unluckily such types of cells have less stability which can be affected by electrodes having liquids. The Redox couple has a corrosive nature and the device can be affected by this nature. However, there are some conditions where solvents can be leaked and an evaporation phenomenon can occur which inhibits the efficiency of this technology [6].

3.1 Structure of PSCs

Organic-inorganic perovskites have astonishing electrical, optical, and magnetic properties, better processability, and tunability of structure which has made researchers curious [7]. Due to these properties and some catalytic impacts, PSCs have great prospects in lasers, fuel cells, and the LED market [1]. The structure of perovskites has a resemblance with ABX_3. In the center, there is a positively charged molecular or atomic cation (having positive charge) of type A. Perovskite cube also has B as cations of atoms at the corner, and the faces of the cube have X as anions of smaller atoms having negative charge as depicted in Fig. 1.

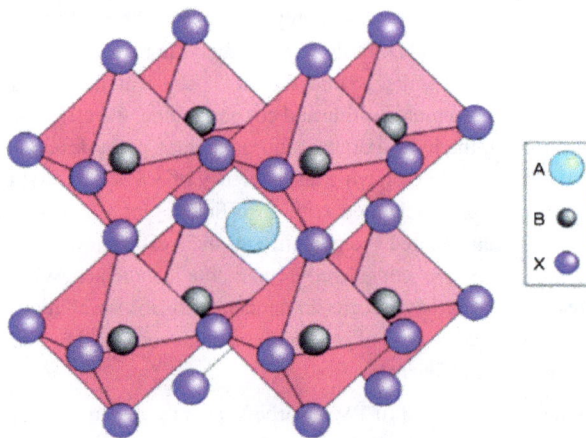

Figure 1. Cubic Crystal structure of Perovskite [1]

The excellent properties like superconductivity, and spintronics of perovskites are achievable if we know what types of atoms or molecules are used in perovskite. In the usual form of ABX_3, the following composition of materials is used in organic-inorganic perovskites.

● A is an organic positive atom/molecule (cation) which is usually methylammonium or cesium.

● B is an inorganic cation that is often lead (Pb^{2+}).

● X is a small halogen (anion) often chlorine and iodine.

A little change in the organic component will affect the inorganic part of the cube because both are interdependent. The Group I V (lead) based PSCs have been most efficient because of good absorption in the visible region and their bandgap is also tunable. There are different reasons that Perovskite materials are preferred over other semiconductors because they have long carrier duration and mobility of modest nature even at low temperatures.

3.2   Optoelectronic Properties Of PSCs

As discussed in the previous section, PSCs have some unique properties that make them so important. Due to the optical absorption, perovskite just requires 500 nm thickness to have significant absorption in the visible region, whereas if we compare it to other solar cells [8], they need to have at least 12μm thickness to absorb significant light to produce charge carriers. The devices which are based on MAPbI$_3$ can have absorption over the whole spectrum till the red region ends, nearly 800 nm which is extendable to 1000 nm by using a Sn-halide combination in devices. Due to this property, perovskites are an important candidate for making tandem devices.

For MAPI$_3$ the carrier diffusion length for both holes and electrons is almost 100 nm and for MAPl$_{3-x}$Cl$_x$ this goes to 1μm, as revealed by PL measurements. For MAPbI3 devices there is a need for the mesoporous electron transporting material (ETM) because holes are more effectively extracted than electrons according to the reports. On the other hand, in Map(I$_3$)$_{2x}$Cl$_x$ based devices there is no need for mesoporous ETM because the carrier diffusion length of holes and electrons already exceeds 1μm.

Instant charge generation is exhibited by MAPbI$_3$, because it is dissociated into high mobility charge carriers of about 25 within 2ps cm^2/Vs. In about 1ps the electron is released into ETL (electron transport layer) of mesoporous TiO$_2$. But unfortunately, due to the intrinsic nature of TiO$_2$, it hinders the mobility of electrons that shall result in unbalanced charge transportation.

To have the complete information about band gap variation, Umari et al. did a complete analysis on density of states. There were some other peaks for MASnI$_3$ that extended from the point of main peak of band gap that were because of 1-p states. These peaks are not present in MAPbI$_3$. There is a rapid increase in levels of valence band (VB) in MAPbI$_3$ if contrasted with MASnI$_3$. In theoretical as well as experimental analysis of the band structure of electrons, spin-orbital coupling is an instrumental factor to be considered. The VB shift in MAPbI$_3$ is about 0.5eV greater than the CB shift which is simply the reason that the band gap of MAPbI$_3$ is superior to MASnI$_3$.

There are some essential optical properties that perovskites possess naming photoluminescence, electroluminescence, and non-linear optical effect [3]. When we use different halides like chlorine and bromine, they also show different optical absorption plus photoluminescence. With adding different metal ions at peak luminescence, the red shift is observed in PSCs.

The electronic properties depend on the inorganic part of the perovskite that further interacts with the organic part due to which orbital overlap changes, which is used for transporting carriers that will affect conductivity. The growth of double and triple layers on perovskite also disturbs electrical conductivity. In 0D, the inorganic part is missing so its conductivity is seemingly lower than 1D and 2D perovskites.

3.3   Influence of A, B, and X site

3.3.1   A-Site

The A-site cation directly impacts properties like the electronic properties due to distortion of B-X bonding which is present in BX_6 octahedra, which arises due to the size effect [9]. The dimensionality for small cations like Cs and MA is 3. When radii keep increasing, the tolerance factor becomes 1, which results in packing symmetry of high order and a reduction in the bandgap.

Methyl ammonium (MA) cation is an extensively utilized cation in different perovskite optoelectronic devices. Due to a small bandgap and high symmetry, larger cations are also useful because they prove to harvest light at enhanced levels. Replacement of MA with different cations has also been tried with other larger cations like Cs but that hasn't worked because large cations alter the dimensionality from 3D to 2D resulting in large bandgap formation.

$MA_xFA_{1-x}PbI_3$ perovskite material containing mixed cations which were introduced by Pellet et al. By optimizing the value of x in these materials the required optical properties are achievable. 14.9% power conversion efficiency was achieved in the current case. Certain regions giving better photocurrent arising from good absorption are responsible for giving better results than MA.

3.3.2   B-Site

Regarding the B site, Pb^{+2} and Sn^{+2} are the most widely used metal cations in organic-inorganic perovskites. Because of superior stability and performance, lead is better than Sn. The stability of different elements in the periodic table starts to reduce on moving from lead to germanium in the table because the inert pair impact is reduced. The counterargument here is that electronegativity starts to increase which results in lowering bandgap and improving conditions for optoelectronic devices. So the possible best option would be an element from lower group IV-A also keeping in view the compromise with the stability of the metal.

The best possible way is the use of $MASnX_3$, which gives the band gap of nearly 1.2eV, but if we use $MAPbX_3$ then it gives nearly 1.5eV, interestingly the first one changes its oxidation state from +2 to +4, because of instant degradation which is caused by exposure of volatile SnI_4 product. Higher symmetry is possessed by Sn-based products, Pb based materials shall have lower symmetry. Tin-based materials have a better possibility of generating photocurrent but rapid use in industry is hindered because of the stability issue that has been discussed earlier.

3.3.3 X-Site

Energy reduction occurs because of a redshift in the absorption of energy which is caused when we move in group VII and the atomic sizes of different halides start to increase. The electronegativity of halides starts to decrease when the atomic size is increased which resembles Pb at some point and gives increased covalent character. The most tunable part of perovskite is the halide part.

Iodine (I): The iodine-based perovskite has high efficiency of about 15% which makes it the most efficient and well-studied material. Pb and iodine together give a stable system because the covalent part of iodine matches well with Pb. Under some perturbation, this becomes oxidized, so this is one of the drawbacks that are noticeable and raises questions about stability issues. Due to this drawback, other options for substitutions of iodine have been examined.

Chlorine (Cl): the most important halide used in perovskite is Cl. Even with so much innovation and enhancement in devices, Cl application in optoelectronic devices is still studied rigorously [10]. The increased carrier life and high diffusion length are possible only due to the high efficiency of Cl-based perovskite materials. Miscibility of iodine with chlorine is higher than with Br due to enhanced efficiency. Cl having a high difference in radii and increased covalent character are the main reasons behind miscibility. It is also seen that mixed halide-based perovskite materials have highly oriented structures while on other hand, while chloride-based materials show simple cubic structures.

Fluoride (F): The property of being highly electronegative and being able to make hydrogen bonding with the other halides could make fluorine a good substitute [11]. But it also has drawbacks of having a big tolerance factor and increased lattice strain that prevents conductivity. The insertion of F in the form of $(BF_4)^-$ that modifies the formula like $MAPbI_{3-x}(BF_4)_x$, could make good hydrogen bonds with MA ions that shall decrease the chance of MA ions volatilization. Because of this inclusion, the conductivity is also lowered. This shows a band gap of about 1.5eV and a tetragonal structure.

4. Mixed Concentration of Perovskite Absorbing Layer

Hybrid organic-inorganic perovskites are very useful in science and technology, because of their tunable optoelectronic properties. The materials can be best described by the general structure like ABX_3 and here B and A are cations and X represents anions.

The highest symmetry is possessed by the ideal perovskite materials, i.e. cubic symmetry. In an ideal case, a Pm3m space group has the highest symmetry in cubic phase perovskite. A ABX_3 has a three-dimensional structure (3D) which is why it is a highly efficient perovskite. Usually, a halide anion motif $X_3 = (I^-, Br^-, Cl^-)$, monovalent cations$(MA, CH_3NH_3, FA, CH(NH_2)_2^+)$ and divalent metal cations $B = (Pb^{2+}, Sn^{2+})$ are mixed [2].

4.1   A-site

The bond gap E_g is affected by changing the bond length and it is influenced by increasing $i.e$ $(FA^+ = 0.19 - 0.22nm)$ or decreasing$i.e$ $(S^+ = 0.167nm, Rb^+ = 0.152nm)$ the size of A. The shape of the perovskite grain is not affected by the concentration of Rb-mixing. But, perovskite film quality deteriorates and after mixing slightly, a blue shift occurs in photoluminescence (PL) peaks [12].

In history, for the first time, an A-mixed cation describing the band gap tunability of $MA_xFA_{1-x}PbI_3$ the based solar cell was proclaimed by Pallel et al. by changing the ratio of MA/FA. In Premixed $MA_xFA_{1-x}I$, the solution was isopropanol which was dipped in pre-deposited PBI_2, so that MA/FA mixed perovskite can be synthesized and that the composition of $MA_{0.6}FA_{0.4}PbI_3$ produced the finest PCE of 13.4-14.52% with extending absorption edge up to 810 nm (Eg $\approx$1.53ev) like Pure $FABI_3$. For the absorption range of the long spectrum enhancement, Lee et al. prepared $FAPBI_3/MAPBI_3$ films with assembled structure by ion exchange and reported the PCE of 16.01% and current density of 20.22 $mAcm^{-2}$. The PCEs of 18.3% resulted in among the optimized MA^+/FA^+ with the composition of $(MA)_{0.6}(FA)_{0.4}PbI_3$ [1,13].

As we already know that inorganic material is more stable than organic material. Therefore, researchers started working on designs that work for inorganic monovalency in the perovskite structure. Choi et al. came up with $CS_x(MA)_{1-x}PbI_3$ PSC with $x = 0.1$ inverted composition: $PSS/Cs_{0.1}(MA)_{0.9}PbI_3/P(BM)A/$ and attained power conversion efficiency of 7.86%. In an enhanced $Cs_x(MA)_{1-x}PbI_3$ cell with $x = 0.9$, higher efficiency was attained reported by Niu et al. Moreover, MAPbI$_3$ has a higher thermal stability than an encapsulated device [12,13].

Rb^+ cations enhance both the performance of Rb-mixed perovskite solar cells and stability, although they have smaller ionic radii than $Cs^+ ions$ (0.181nm). Park et al. examined (FA/Rb) PbI$_3$ and attained power conversion efficiency of 16.15% by using the same Rb quantity. Rb-mixed also got attention, because it enhanced stability against moisture at 85% RH [12].

4.4 Mixed B-Sites Cations

Large-scale application of Pb is restricted due to its toxic nature and therefore lead free alternatives were sought after. As in the periodic table, Sn and Pb belong to the same family. Therefore, Sn takes the place of Pb in the primary stage. The perovskite material is made with Sn^{2+} has a lower band gap than the perovskite material made of Pb^{2+}. Hence, to attain absorption near the infrared region we prepared perovskite material with mixed Sn and Pb [1,13].

It was found that it is possible to adjust the band gap within the range of $1.17eV - 1.55eV$. When Ogomi studied the optical properties of $MASn_{1-x}Pb_xI_3$ with varying ratios of Pb to Sn, the optical absorption wavelength was increased up to 1070 nm. A decrease in open-circuit voltage was noted due to the narrow band gap. In this whole process, lower power conversion efficiency was achieved because Sn^{2+} was easily oxidized to Sn^{4+}. In Sn-based perovskite solar cells, the conventional plane gave PCE of $5 - 6\%$. Zou et al. in an inverted planer heterojunction got PCE of 10.1% by synthesis of binary Pb-Sn perovskite $(MAPb_{1-x}Sn_xI_{1-y}Cl_y)$ [13].

An impressive PCE of 13.9% was attained by Liu et al. with C$_{60}$-modified Sn-Pb perovskite film $(MAPb_{1-x}Sn_xI_3)$. Hybrid Sn-Pb perovskite cell gives good performance and stability when disclosed in surroundings in the absence of encapsulation after perovskite solar cell modified with C$_{60}$. For improving stability, Marshall explained the addition of $SnCl_2$ in the light-absorbing layer, but it could not affect energy conversion efficiency [13].

4.5 X-Site

When we move down in the group of halogen atoms, the size of halides surges, creating the redshift of energy absorption. Mostly this occurs because the electronegativity decreases when the size of the atom increases and at indefinite time matches with Pb hence giving extra covalent character. The admirable studied material in the perovskite solar cell is iodide and iodide-based perovskite material gives efficiency as high as 16%. Iodide along with lead gives a very stable system because it matches nicely with Pb having a similarly covalent structure. But, the issue arises here because iodide is easily oxidized [1].

The most popular halide used in perovskite material is chlorine. As chlorine enhances diffusion length and carrier lifetime resulting in increased efficiency. Another fascinating effect of chlorine-doped halide perovskite is having fewer miscibility properties compared to bromine and iodine. We see this behavior because bromine and Iodine have smaller degrees of covalent character. It has been noted that $MAPbI_{3-x}Cl_x$ and $MAPbI_3$ with a maximum 3 - 4% Cl to I ratio has similar stoichiometry [1].

5. Requirements for Each Layer

5.1 Electron Transport Layer

PSCs are highly dependent on the type of material that is selected for the electron transport layer (ETL) in the context of efficiency and stability. The main property of ETL is that it should fulfill the band symmetry with the active layer. Transmittance of the ETL should be high so that photons can easily pass and can be absorbed by the active layer [14]. P-i-n and n-i-p are two types of devices based on the arrangement of the electron transport layer or hole transport layer (HTL). Due to the structure of ETL, n-i-p type PSC is classified into mesoporous or planar configuration. There are five types of layers present in both types of devices i.e., (i) TCO (transparent conducting device) such as (fluorine-doped tin oxide) FTO and (indium tin oxide) ITO (ii) electron transport layer (ETL) (iii) light absorbing layer (iv) hole transport material (HTM) (v) metal or non-metal electrode [15].

The presence of ETL is vital for the high performance of PSC, for maximum electron collection and for transferring electrons from the perovskite layer to the electrode. Some groups claimed that without using ETL, they achieved 14% PCEs but the ETL layer in PSCs was more effective in terms of stability and performance. Jurez-Perez *et al.* noted that the absence of ETL slightly affects V_{OC}(Open-Circuit Voltage) but strongly influences J_{SC}(short circuit current density) and critically reduces it. The absence of ETL also reduces the fill factor which has a direct influence on the power conversion efficiency of the devices. Electron-hole recombination increases in the presence of ETL at the small cost of an increase in series resistances. Power conversion efficiency decreases from 14.1 to 11.5% after the removal of ETL as observed by Zhang and co-workers. Ke et al. reported that power conversion efficiency was 14% but complete cells showed power conversion efficiency of 16% and there was no sign of stabilized power observed in the absence of ETL, however, PCE achieved from this current-voltage measurement was a little high [15].

To manufacture high-efficiency perovskite solar cells, the nominated ETL material must satisfy several criteria such as (i) Electrons must move fast within ETL so electron mobility of material must be decent (ii) To ensure high transparency, the band gap of incident light passing through material should be wide (iii) Compatible energy level. Further in ETL planar heterojunction, the semiconductor material selected must have (i) a good anti-reflection to lessen the transmittance (ii) a compact structure to prevent direct contact between transparent conducting oxidize and perovskite. Generally, two ways are used to improve the properties of electron transport layers (i) modifying the structure of ETL and (ii) using a material with better optoelectronic properties [15].

5.1.1 Different ETL Material Used In Perovskite Cells

TiO_2, ZnO, SnO_2 and ZrO_2 are some metal oxides that are commonly used as ETL. Each of these materials has its own benefits to enhance power conversion efficiency. Two types of processes are introduced to fabricate perovskite solar cells. One is based on one step-deposition, which gives a PCE of 7.23% in comparison to a TiO_2 based perovskite solar cell. To control shunt resistance, we need optimum porosity over the layer of perovskite [14].

Yella et al. modified the particle size of rutile TiO_2 successfully by using chemical bath-deposited rutile thin film to give PCE as high as 13.7%. Luo et al. explained the complete list of hole injection, electron injection, and exaction annihilation. It is very difficult to define the characteristics of an ideal ETL due to the large variety of ETL/ETM. Using SiO_2, Al_2O_3 and ZrO_2(as scaffold material), high efficiency was reported. A research group reported that they achieved ~10% PCEs by using SiO_2, Al_2O_3 and ZrO_2. We can successfully form a perovskite layer, block holes and inject electrons, when the CB edge (CBE) is lower and the VB edge (VBE) is higher, which can improve the recombination resistance. Smooth electron injection is observed in both SiO_2 and ZrO_2, because for SiO_2 and ZrO_2, the much higher CBE and band gap is large, as a result electrons stay in the CB of the perovskite layer for a much longer time [14].

Bi et al. described that ZrO_2 had a higher efficiency than TiO_2 after a comparison of both the scaffold materials was drawn. ZrO_2 had PCE = 10.8% and TiO_2 had PCEs = 9.8%. ZrO_2 is an efficient scaffold material because of higher photo-voltage and electron lifetimes, although the band alignment was not fulfilled [14]. It has been reported in the literature that using organic semiconductors, metal oxide-free perovskite solar cells have been achieved. In order to get lofty efficiency in inverted PSC organic ETL such as [6, 6]-phenyl C61-butyric acid methyl ester (PCBM), poly (3-hexylthiophene) (PEHT), PEHT: PCBM composites. PSC can be made with enhanced performance and efficiency by using organic material with a metal oxide scaffold blocking layer. The LUMO of such materials should be lower than that of the active layer, this condition is also satisfied by many organic materials. An ETL was made by NiO_2, scaffold blocking layer and HTL was made with $PCBM/BCP$ as reported by Wang et al. $Furthermore,$ 16.21% efficiency was attained with an extraordinarily high $V_{oc} = 1.05V$ and $FF = 0.78$ with planar heterojunction device constructed as $ITO/PEDOT:PSS$/Perovskite/$PC_{61}BM/Cu/Al$ reported by Chiang et al. [14].

5.2 Hole Transporting Layer

The hole transport layer is basically used for three purposes. First, to avoid the straight exposure of metal electrode with mesoporous TiO_2 perovskite, it is placed prior to the gold electrode, as a result, V_{oc} is increased and greater luminesce performance is achieved because recombination is reduced. Secondly, by decreasing the (diffusion) losses of charges in HTL, the illumination wavelength is increased and the central quantum efficiency is independent of applied voltage. Third, allowing an alternate way for light to pass between the absorber layer, the reflectivity of the gold electrode is increased due to hole transporting layer. These effects enhance short circuit current density [16].

5.2.1 Hole Transporting Material (HTM)

Polymeric, inorganic, and small organic HTMs are basic categories in which solid-state hole transporting materials are divided. Inorganic materials have high hole mobility, low cost, and good properties, hence these are used in PSCs, such as NiO, CuI, and CuSCN. But the stability of devices based on these materials is affected. Complex purification processes, poor infiltration into nanostructured material, low solubility, and tricky characterization are some drawbacks of polymeric materials [17].

5.2.2 Inorganic P-type semiconductors as HTMs

HTL inorganic materials were investigated due to their large hole mobility, low-cost construction, and intrinsically high stability. However, there is a small drawback as in mesoscopic PSCs, the solvent utilized for deposition dissolves within the perovskite. Kamat and co-worker attained a PCE of 6% for copper iodide (CuI). Due to low V_{oc}, poor PCE was achieved, while the highest value of fill factor (F) (0.77) was attained, which can be credited to high recombination measured by impedance spectroscopy. Copper thiocyanate (CuSCN) being transparent to the infrared and visible region, and possessing high mobility was used as HTM initially by Ito et al. into a mesoscopic PSC. CuSCN under standard AM1.5G irradiation gives a maximum PCE of 86% when deposited by a doctor-blade process. PCEs of 12.4% was recorded after optimization of the thickness of the hole transport material (600-700nm). Because of effective charge removal from perovskite and by using CuSCN in a planar structure perovskite solar cell, a high J_{sc} ($19.7mAcm^{-2}$) was achieved. Bian and Liu et al. achieved an impressive PCE of 16.6%. HTM up to 57nm of thickness by electrode position and $MAPBI_3$ formation was used in a one-step fast deposition crystallization method. In DSSC and OPV, nickel oxide was utilized as hole transporting material, and a power conversion efficiency of 7.26% was achieved by Sarkar et al. by using a layer of NiO placed by electrode position in an inverter planar structure [17].

5.2.3 Organometallic HTMs

In mesoscopic PSCs, CuMePc (tetramethyl- substituted Cu" phthalocyanine) was utilized as a hole transporting material and it enhances the efficiency in comparison to unsubstituted phthalocyanine. The reported PCE was only 5%; although it was hypothesized that methyl substitution would increase the $\pi - \pi$ interaction. The molecular construction of hole transporting material commonly contains some electron-rich groups conjugated to the inner aromatic core cell called HTM core. Triazines, thiophenes, diketopyrrolopyrrole, triarylamines and spiro-liked HTMs are some examples of conjugates that have been reported. By changing $P - OMe$ with several alkyloxy groups, many Spiro-OMeTAD analogues were constructed. Spiro-OMeTAD is of a high-glass transition temperature T_G and has exact energy level and in PSCs, it is still more effective as several hole transporting materials. Hole-mobility, μ_b of pristine Spiro-OMeTAD was decided by (SCLC) space charge limited current and it was about $2 \times 10^{-4} cm^2 V^{-1} S^{-1}$ [17].

5.3   Absorbing Layer

In a perovskite solar cell, an absorber layer is a perovskite layer that provides free electrons and holes after absorbing light. Upon the action of an electric field, these free charges diffuse and drift away. In short, electrons move towards HTL and holes move toward the ETL. The ability of transporting holes and electrons toward their respective layers gives the efficiency of the absorbing layer. The electron dissipates energy, when holes and electrons are collected by their specific electrode, and before returning to the device at the opposite side, electron, and hole recombination occurs. It will recombine with the hole if electrons are not extracted by the electron transport layer. This is nothing but a radiative transition, in which a photon of energy equal to the band gap is emitted.

5.3.1 Preparation Method of The Perovskite Light Absorbing Layer

Roughly there are three main types of synthesis processes commonly used for light-absorbing layers of perovskite solar cells such as the vapor-assisted solution method, vapor-deposition method, and the solution method. In the vapor-assisted solution method, the HTL directly touches the electron transport layer and produces a synthetic crystal, thus the open circuit voltage and filling factor are reduced. But, the solution method is easy and inexpensive. Using vapor deposition method, the perovskite film so formed has a high surface density and hence enhances the open circuit voltage as well as the filling factor. However, this process involves high energy consumption and a high vacuum is required. If we integrate the advantages of solution and evaporation methods it is known as the vapor-assisted solution method [18].

6.   Fabrication Techniques

The quality of the absorber layer film affects mainly the performance of other layers of PSCs exhibiting equal performance. To find the quality of the film, surface morphology of the substrate being used is preferred for the fabrication from the several factors of high quality film. To get good device performance and high quality film, the substrate being used is usually mesoporous. The various deposition techniques for the construction of PSCs are 1-step deposition, 2-step deposition, vapor deposition method, spin coating, and TVP (thermal vapor deposition) [2].

6.1   One-Step Deposition

1-step deposition method has been mostly used by scientists as it is a reliable process and cheaper. Perovskite precursor and pinhole-free is the technique used in the fabrication of perovskite film. In a single-step deposition, the organic solvent is mixed with the following; "MX_2 (M is Pb or Sn and X is Br and Cl) mixture and formamidinium iodide (FAI) or methylammonium iodide (MAI)" as the perovskite precursor. This mixing of MAI or FAI (organic halide) and MX_2 (Inorganic halide) in an organic solvent, a pure phase, pinhole-free, and a dense perovskite is obtained by spin coating and tempering in the range of 100-150 ^{0}C [19]. A one-step deposition is preferred due to slow crystallization in the fabrication process. Another factor One-step deposition gives us is that these are some conditions that determine the quality of perovskite film. The conditions are tempering, humidity, the substrate being used, and as well as film thickness. The two parameters that are important for a good-quality film are film thickness and morphology [20].

6.2   Two-Step Deposition

In order to expose organic ions in perovskite film, whether the ions are in solution or in vapors, PbI_2 is spin-coated on the substrate [21]. After this spin coating of lead oxide (PbI_2), there is a dipping of the substrate into an isopropanol solution of methyl aluminum iodide (MAI). For the formation of perovskite having organic and inorganic metal halide, PbI_2 is again spin-coated onto the MAI. This method is being used for the uniform and controlled fabrication of thin film in the PSCs [1]. With a certified PCE of 14.14%, two-step deposition first showed 15% of PCE. In this method, PbI_2 is not completely converted, and also crystal size is not controlled with surface morphology. So different post-operative methods have been revealed [21].

In one-step deposition, a comparatively smaller film thickness of 250 nm is obtained while it is found that two-step deposition leads to a film thickness of approximately 280 nm. Field emission scanning electron microscope (FESEM) gives the surface morphology of the film as illustrated in fig. 2. While distinct perovskite grains are obtained by heating the precursor solution beforehand. In this method, a film having a large number of pinholes is obtained. ETL and FTL are directly influenced by the pinhole presence that results in the increase in reconnecting of charges in the PSC. The basic substrate is totally covered by the condensation of PbI_2 film via deposition in a steamy atmosphere. All the characteristics ensure the formation of a condensed perovskite layer via the beforehand deposition of PbI_2 film which reacts with the MAI solution; also it is found that TSD fabrication techniques give a mountain-like structure with less smoothness. On comparing the grain sizes of both OSD and TSD, it was noted that the mean grain size of approximately 220 nm (TSD) was comparatively greater than the grain size of 90 nm (OSD). Condensed structure and greater grain size were the two characteristics that determine good performance by finer transfer of charges in two-step fabricated technique. Such fabricated structures made certain the full separation of ETL and HTL and averted critical charge reconnection, it is notable that pinhole-free condensed structure was more important than grain size in determining the device performance. Another factor of the morphological analysis showed that TSD was more beneficial for high steamy processing in perovskites [9].

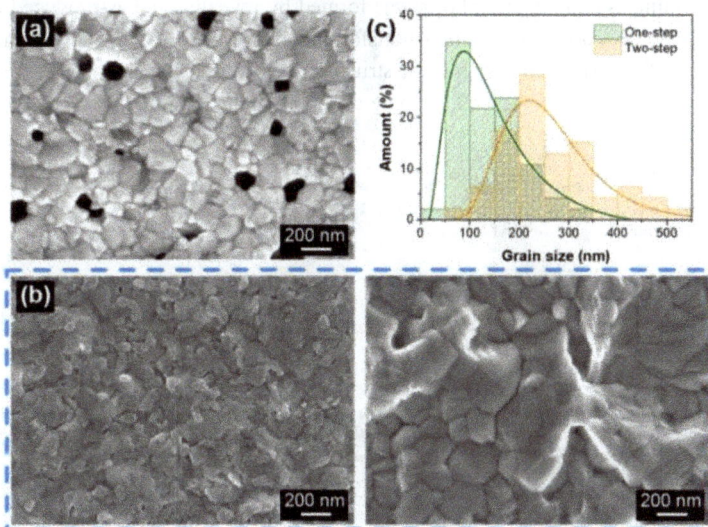

Figure 2. *FESEM pictures of surface morphology of (a) Perovskite film synthesized through OSD, (b) PbI2 film (left) and perovskite film (right) synthesized through TSD in high-steamy (RH40-50%). (c) Dispersal of perovskite grain size [9].*

6.3   Vapor Deposition Method

Out of the techniques mentioned above, the vapor deposition technique has used for large-scale production in the semiconductor industry. It has vast applications in optoelectronics. Vacuum-vapor deposition was first proclaimed by Salau et al. to determine the perovskite film [21]. In this method, the top of the substrate is coated with a layer of PbI_2. There is no solution-dipping process in this technique. After the coating of PbI_2, a vapor deposition technique is applied and MAI is then deposited onto it.

During the growth process of perovskite, thermodynamic stability and kinetic reactivity of MAI are mainly focused to make films having definite grain structure, big size of crystal, full surface coverage and less roughness of surface for Photovoltaic (PV) applications. This method gives PCE of 12.1% which is the highest power conversion efficiency in planar structure [20]. On the FTO/TiO_2 substrate, 150nm PbI_2 thin film was first vaporized and then for 30 min kept in a MAI vapor environment under 180°C. The SEM image reflects that $MAPbI_3$ is composed of condensed grain size of approximately 800 nm. In the BAI vapor, a 2-dimensional capping layer was formed by revealing the as-produced $MAPbI_3$ thin film. Fig. 3. shows the limited transformation of 3-dimensional to 2-dimensional structure by the formation of apparent structure in 5 minutes from the 3D grains. The transformation of morphology takes some time and after 60 minutes by reaction, a low level structure was noticed [9].

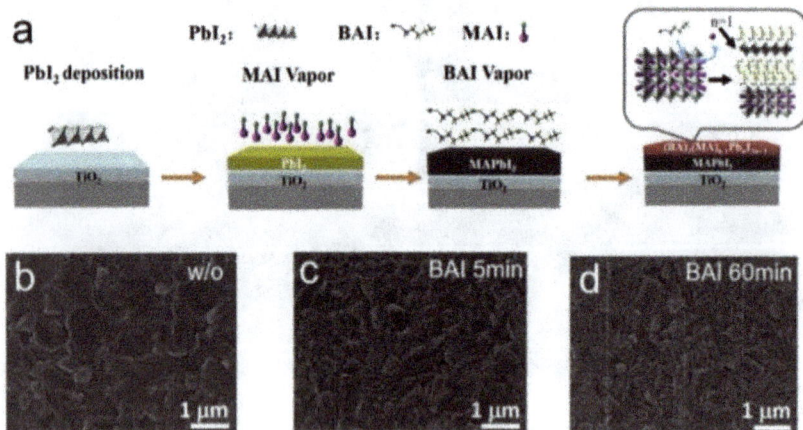

Figure 3. Schematic illustration of VDM [8]

6.4 Spin Coating

Amid the solution processing based techniques, spin coating is easy and economical. It gives a uniform deposition of the perovskite layer of PSCs. In this method, two types of structures are shown; one is inverted PSC and the second is regular PSC [22]. This method is used for making small-sized devices with an area of 0.1 cm^2 and large PSCs of 1 cm^2. A perovskite film fabricated through spin coating has a PCE of more than 9.4%. By handling the crystal growth and direct hardening time, the quality of the film can be increased. The fabrication through spin coating followed two processes as follows.

6.4.1 One-Step Spin Coating

Methyl aluminum iodide (MAI) and formamidinium iodide (FAI) are organic halides salts with lead iodide, lead bromide (lead halide salts) and then dissolved in dimethyl sulfide (DMSO) or N-dimethylformamide (DMF) as solvent. Then, this precursor solution is coated on a movable electron, a hole-carrying layer, and absorber layer. After the tempering of this layer, a perovskite film is formed [23].

6.4.2 Two-Step Spin Coating

In the current technique, halide organic salt is mixed in dimethylformamide and lead halide salts are mixed with isopropanal to be coated on the ETL/HTL. Next, this solution after the coating is annealed on a lead halide surface, there is a spin coating of lead halide solution and tempering by inner spreading to get perovskite films. An experimental observation by Jiang et al. on this technique showed 20.1% PCE in 1 cm^2 PSCs [23].

The film width and quality can be improved by fixing time of spin coating, speed of spin and acceleration of spin. Top recorded efficiency using this technique is 22.1% at laboratory scale. Two structures i,e; regular and inverted PSC can be obtained using this technique of fabrication. Although efficiency obtained is high, this technique is not being used for producing films on larger areas. In case of bigger substrates, film thickness also varies from center to end [22].

The preparation method of this process is not so complicated and the attractiveness and width of perovskite film can be controlled. DMSO and DMF are precursor solutions that consist of polar aprotic solvents. At high revolutions per minute, the substrate is spun so that solvents can be abolished by centrifugal force [24].

6.5 Thermal Vapor Deposition

David B. Mitzi first gave the concept of thermal vapor deposition of the perovskite absorber layer. After this, Liu et al. introduced the scope of this technique by giving the PCE of

15.4% with the help of $MAPbI_{3-x}Cl_x$ PSCs in the form of planar heterojunction. In this method, a glove box filled with nitrogen is used to deposit the $MAPbI_3$ absorber with the use of an evaporation system having two sources. Over the FTO-coated glass, an initial deposition of a compact layer is formed. On the top of the compact layer, perovskite films having vapors are then fabricated. This vapor deposition showed some reactions which are critical to control [20].

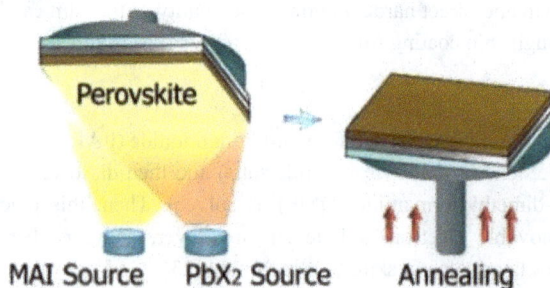

Figure 4. Schematic diagram of TVD process [20]

In this method, the area of the film is totally converged. At high vacuum, there is an equal evaporation of MAX and PbI_2 at the same time with certain molar ratios. After this evaporation, a reddish brown film was obtained and then this film was tempered. In the nitrogen-filled glove box tempering perovskite films provides the following attributes; dark color, complete crystallization, and outward growth of the crystal. By completely covering the substrate, the conversion of $MAPbI_3$ into varied morphologies with equally contributed perovskite polycrystals having sizes between 100 and 200 nm and distinct platforms are achieved. The results of making $MAPbI_3$ crystalline film on spin-coated films, were reported by Yu et al. [3].

7.   Challenges in Perovskite-Based Solar Cells

7.1   Stability Challenges

Owed to achieving great efficiency in a relatively small period of time, the perovskites based solar cells (PSCs) are becoming powerful energy sources for the future. But the photovoltaic cell's stability throughout a long period is the biggest challenge that prevents the perovskites from being commercialized rapidly, and this factor leads to the degradation of PSCs [25]. The instability of PSCs is the biggest issue in manufacturing of the perovskite solar cells because during the characterization of the cell, its phase changes rapidly [26].

There are four factors that are identified that are responsible for the degradation of PSCs, including temperature, UV light, oxygen, and moisture [27]. The significant degradation of the PSCs takes place in ambient air due to the presence of water and oxygen which affects its chemical stability, and it gives rise to a transition from dark brown to yellow in appearance [28]. This issue resists the PSCs from being used for outdoor applications.

7.2   Thermal Effect

Thermal instability is also the biggest challenge to the manufacturing of PSCs-based technologies. PSCs that have direct exposure to sunlight will experience an upsurge in the temperature of the solar panel. For solar cells, the ambient temperature is 40°C and heat accumulation can result in a temperature of up to 85°C [29]. The thermal stability of $MApbI_{3-x}Cl_x$ was investigated and discovered that the degradation occurs at 85°C, which implies that perovskite solar cells may not be employed during daytime as the device temperature exceeds to 85°C [30]. To understand the thermal degradation of PSCs, calculations were made according to which the formation energy of $MApbI_{3-x}Cl_x$ was 0.11-0.14 eV [6], that is almost approachable to 'thermal energy (0.093)' predicted at 85°C which implies that degradation occurs at 85°C [31]. To make use of highly thermally conductive materials, the instability due to temperature can be reduced by decreasing heat accumulation inside the PSCs. According to Conings et al., the perovskite layer decomposes into lead iodide and organic components, then is completely removed by heating it at 85°C for 24 hours in the dark. The thermal stability can be improved in this way [32].

7.3   Toxicity

PSCs have a certified efficiency of more than 22% for photovoltaic applications, which has recently gained major attention. But unfortunately, the presence of lead in perovskites materials raises serious concern about their future commercialization [33]. The toxicity of Pb is another big challenge towards the development of perovskite based technologies. Although the Sn and Ge have been offered as an alternative component for the lead, the issue is the low efficiency of the PSCs [26]. Due to toxicity of lead, and thus its potential hazard on human health and the ecosystem, they have not commercialized. To rectify this problem, scientists recommended two corrective measurements. First one is to substitute the lead based PSCs with a less toxic material or total removal of lead by replacing this with metallic material (tin, germanium, copper, bismuth, or antimony). However, it decreases the overall efficiency of the perovskites based solar cells [34]. Thus, the new candidate to replace the lead must exhibit high potential with respect to performance, cost-effectiveness, and stability [35].

7.4   J-V Hysteresis

Hysteresis is a fundamental issue for the commercialization of PSCs. The J-V curves of the PCs are to be affected by several aspects of the measurements, scan direction, rate, and range as well as voltage treatment. The conventional solar cells do not alter the nature of cadmium telluride and silicon during the scanning of forward bias voltage to reverse bias, but the PSCs exhibit an anomalous J-V hysteresis. The factors on which the J-V hysteresis depends are ferro electricity [36], unbalanced charge collection and ion migration [37]. The hysteresis-free solar cells give rise to a significant current density-voltage curve as when the temperature of the cell is dropped to 175K [38]. Numerous ways are used to overcome this J-V hysteresis including surface passivation or boosting the PbI_2 content. The use of mesoporous TiO_2 significantly reduces the hysteresis effect, whereas similar planner devices exhibit significant hysteresis because of the induced dipole polarization into the perovskites. By the implication of thiophene or pyridine treatment, we can minimize the hysteresis effects in planner devices by passivation of the perovskite layer's surface [39].

Various types of techniques are described for resolving these challenges and increasing the efficiency of perovskite, where they can be used commercially. From this discussion, we come to know that PSCs are not dependent on the device's architecture, but by enhancing the stability of the perovskite, we can decrease the hysteresis effect.

8.   Efficiency of Perovskite

The solar cell's efficiency is measured by using Eq. 1,

$$J_{SC} = \frac{I_{SC}}{A} \tag{Eq.1}$$

Where;

J_{SC} = current density of short-circuit

And, A = the solar cell's active area

The PCE (power conversion efficiency) is to be expressed by η, given by the Eq.2,

$$\eta = \frac{FFV_{OC}J_{SC}}{P_{OP}} \tag{Eq.2}$$

Where,

P_{OP} = Optical power density

And FF indicates the fill factor, and it is stated simply as the factor which defines a solar cell's maximum power output. The mathematical expression of the fill factor is given by Eq.3,

$$FF = \frac{P_{max}}{V_{oc}I_{sc}}$$ (Eq.3)

Where V_{OC} is stated as the voltage developed across the photovoltaic cell's two open terminals. And I_{SC} indicates the current of the short circuit.

The organo-metallic halide perovskite is being explored extensively for high-efficiency solar cells, utilized as a light absorber for PSCs. Through worldwide efforts, a gain in efficiency of the PSCs up to 25.2% has been reported recently. The perovskite LED's electroluminescence efficiency has improved considerably in a relatively brief time frame, reaching a maximum EQE (external quantum efficiency) of above 20 %, that is analogous to the luminescence of phosphorescent materials [40]. Inorganic halide perovskite ($CH_3NH_3PbX_3$, where X = chlorine and bromine) has been used to make solar cells which is a low-cost next-generation solution process able material, having reached to a recognized PCE of 25.2% [41].

By optimizing the device's structure of perovskite solar cell, the PCE of the PSCs increases from 3.8 % to a score of 25.2 % which is equivalent to a single crystal solar cell. It may be stated that efficiency is no longer the tentative block to PSCs commercialization. The instability of the PSCs is a big issue, by overcoming this issue, the efficiency of the PSCs can also be enhanced. The polymer methods used have significant results in increasing device stability. As a result, despite improving the stability of PSCs, this device has achieved a PCE of up to 21.6% with no current density-voltage hysteresis [41].

In 2011, Jeong developed perovskite quantum dot photovoltaic cells with the EQE (external quantum efficiency) of 79% and the PCE of 6.54%. During 2012, Kim and his coworkers devised power conversion efficiency of 9.7% of $CH_3NH_3PbI_3$ based PSCs [42]. In 2013, it was discovered that by lowering the temperature of $CH_3NH_3PbI_{3-x}Cl_x$ solar cells from $500\ to\ 150°C$ produced PSCs, the PCE of 12.3% was achieved for this perovskite [43]. The 15% power conversion efficiency of solid-state PbI_2 PSCs was attained, when Burschka et al. manufactured the photovoltaic cell by using a two step deposition method [44]. In the history of solar cells, formamidinium lead trihalide was shown to have a PCE of 14%, making it a plausible choice. Furthermore, in 2015 Yang and his colleagues replaced the previously utilized NH_3PbI_3 with $FAPbI_3$, and the photovoltaic cells that were based on $FAPbI_3$ were found to have high absorption band gaps, with PCE of up to 20.1% [44].

Moreover, Sheng et al. created a thin film of $CH_3NH_3PbBr_3$ PSCs by vaporizing a $CH_3NH_3PbBr_3$ layer onto TiO_2. The photoluminescence quenching technique yielded a 1.06 m diffusion length and 8.7% of conversion efficiency [45]. Cho et al. enhanced the solar cells' efficiency for photovoltaic devices by loading dichalcogenides nanomaterial onto a highly porous TiO_2 surface that contains a hole transporting substance between both electrochemical cells [46].

9. Future Perspectives

Organic-inorganic hybrid compounds have seen a notable rise in popularity due to the promise of new technical advancements they hold in the electronic age we're entering. PSCs are still under research, and the ideal operating conditions are unknown, but researchers have been endeavoring to disappear these obstacles just to widen the availability of this technique. PSCs are hoped to be powered by solar energy, in near future. It is possible that tandem cells, a photovoltaic device that combines silicon-based perovskite cells, will emerge as a more powerful photovoltaic technology. Even though PSCs are very inexpensive and abundantly available, their stability over a long time, toxic effect, and disintegration in a wet climate must be all significantly reduced. Researchers have focused their efforts on overcoming these limitations and providing meaningful information that can help to resolve the most pressing problems. As an example, researchers have developed lead-free PV cells that are more stable than lead-based perovskite cells which represent a significant threat to the environment. It is also important to note that new ways are being developed to stabilize perovskite materials, which have the potential to increase not only the performance of solar cells as well as other perovskite-based products including photodiodes and LEDs.

The world economic forum claims that perovskite developed as ink can be used to print on anything, and alternatively, it could be rolled down to textiles and materials that are woven into cloth or used as construction components. In addition, the worldwide PSCs market is analyzing profit, fragmentation and pricing, and growth in output for PSCs to be commercialized in near future. Light-harvesting layers that can be manufactured in large quantities open the door to a rise in the market of renewable energy sources

Conclusion

Energy crises have been a prevalent problem for a few years and world researchers are constantly working to derive new energy sources and improve the existing ones. Silicon-

Materials Research Forum LLC
https://doi.org/10.21741/9781644902738-1

based solar cells have ruled the photovoltaic industry, but they are expensive. Perovskites, especially hybrid organic-inorganic perovskites are the new contenders in the photovoltaic industry, which promise cost-effectiveness and higher efficiency. Perovskites-based solar cells possess great optical and electronic properties which are influenced by modifications in A, B, or X sites, which provide us ways to optimize their performance. The basic components of PSC such as the electron transport layer, hole transport layer, and absorber layer are deliberated along with materials that are being used for these layers. The structure of the perovskite film and morphology are the two factors that are basically controlled by deposition techniques. Different fabrication techniques such as one-step deposition, two-step deposition, spin coating, and vapor deposition are discussed and compared. Perovskite solar cells have become significantly popular in recent years, paving the way for new technological advances in photovoltaic technologies. This field's developments are expected to have a significant impact on future economic growth. The main issues with these perovskite solar cells are their toxicity and cell disintegration and hence efforts are being made to minimize these factors

References

[1] M.K. Assadi, S. Bakhoda, R. Saidur, H. Hanaei, Recent progress in perovskite solar cells, Renew. Sust. Energ. Rev. 81 (2018) 2812-2822. https://doi.org/10.1016/j.rser.2017.06.088

[2] J.M. Ball, M.M. Lee, A. Hey, H.J. Snaith, Low-temperature processed meso-superstructured to thin-film perovskite solar cells, Energy & Enviro. Sci. 6 (2013) 1739-1743. https://doi.org/10.1039/c3ee40810h

[3] D. Banerjee, K.K. Chattopadhyay, Hybrid inorganic organic perovskites: A low-cost-efficient optoelectronic material, in: S. Thomas, A. Thankappan (Eds.), Perovskite Photovoltaics, Elsevier, 2018, pp. 123-162. https://doi.org/10.1016/B978-0-12-812915-9.00005-8

[4] K.P. Bhandari, R.J. Ellingson, An overview of hybrid organic-inorganic metal halide perovskite solar cells, in: T.M. Letcher, V.M. Fthenakis (Eds.), A Comprehensive Guide to Solar Energy Systems, Elsevier, 2018, pp. 233-254. https://doi.org/10.1016/B978-0-12-811479-7.00011-7

[5] P.P. Boix, K. Nonomura, N. Mathews, S.G. Mhaisalkar, Current progress and future perspectives for organic/inorganic perovskite solar cells, Mater. Today. 17 (2014) 16-23. https://doi.org/10.1016/j.mattod.2013.12.002

[6] P.P. Boix, K. Nonomura, N. Mathews, S.G. Mhaisalkar, Current progress and future perspectives for organic/inorganic perovskite solar cells, Mater. Today. 17 (2014) 5-20. https://doi.org/10.1016/j.mattod.2013.12.002

[7] J. Burschka, N. Pellet, S.-J. Moon, R.H. Baker, P. Gao, M.K. Nazeeruddin, M. Grätzel, Sequential deposition as a route to high-performance perovskite-sensitized solar cells, Nature. 499 (2013) 316-319. https://doi.org/10.1038/nature12340

[8] L. Calió, S. Kazim, M. Grätzel, S. Ahmad, Hole-transport materials for perovskite solar cells, Angew. Chem. Int. Ed. 55 (2016) 14522-14545. https://doi.org/10.1002/anie.201601757

[9] B. Chaudhary, T.M. Koh, B. Febriansyah, A. Bruno, N. Mathews, S.G. Mhaisalkar, C. Soci, Mixed-dimensional Naphthyl Methyl Ammonium-Methylammonium Lead Iodide perovskites with improved thermal stability, Sci. Repo. 10 (2020) 1-11. https://doi.org/10.1038/s41598-019-56847-4

[10] C.-H. Chiang, Z.-L. Tseng, C.-G. Wu, Planar heterojunction perovskite/PC71BM solar cells with enhanced open-circuit voltage via a (2/1)-step spin-coating process, J. Mater. Chem: A, 2 (2014) 15897-15903. https://doi.org/10.1039/C4TA03674C

[11] D.H. Cho, H.W. Choi, Fabrication and characterization of mixed Lead halide thin films for perovskite solar cells, Mol. Cryst. Liq. Cryst. 654 (2017) 201-208. https://doi.org/10.1080/15421406.2017.1358045

[12] B. Conings, J. Drijkoningen, N. Gauquelin, A. Babayigit, J. D'Haen, L. D'Olieslaeger, A. Ethirajan, J. Verbeeck, J. Manca, E. Mosconi, Intrinsic thermal instability of Methylammonium Lead trihalide perovskite, Adv. Energy Mater. 5 (2015) 1500400-1500477. https://doi.org/10.1002/aenm.201500477

[13] V. D'innocenzo, G. Grancini, M.J. Alcocer, A.R.S. Kandada, S.D. Stranks, M.M. Lee, G. Lanzani, H.J. Snaith, A. Petrozza, Excitons versus free charges in organo-Lead trihalide perovskites, Nature Commun. 5 (2014) 1-6. https://doi.org/10.1038/ncomms4586

[14] A. Dualeh, P. Gao, S.I. Seok, M.K. Nazeeruddin, M. Grätzel, Thermal behavior of Methylammonium Lead-trihalide perovskite photovoltaic light harvesters, Chem. Mater. 26 (2014) 6160-6164. https://doi.org/10.1021/cm502468k

[15] G.E. Eperon, S.D. Stranks, C. Menelaou, M.B. Johnston, L.M. Herz, H.J.J.E. Snaith, E. Science, Formamidinium Lead trihalide: A broadly tunable perovskite for efficient planar heterojunction solar cells, Energy & Environ. Sci. 7 (2014) 982-988. https://doi.org/10.1039/c3ee43822h

[16] J.M. Frost, A. Walsh, What is moving in hybrid halide perovskite solar cells? Accounts of chemical research, ACS Pub. 49 (2016) 28-535. https://doi.org/10.1021/acs.accounts.5b00431

[17] S.L. Hamukwaya, H. Hao, Z. Zhao, J. Dong, T. Zhong, J. Xing, L. Hao, M.M. Mashingaidze, A review of recent developments in preparation methods for large-area perovskite solar cells, Coatings. 12 (2022) 250- 252. https://doi.org/10.3390/coatings12020252

[18] J.-H. Im, H.-S. Kim, N.-G. Park, Morphology-photovoltaic property correlation in perovskite solar cells: One-step versus two-step deposition of CH3NH3PbI3, Apl. Mater. 2 (2014) 081497- 081510. https://doi.org/10.1063/1.4891275

[19] M. Jamal, M. Bashar, A.M. Hasan, Z.A. Almutairi, H.F. Alharbi, N.H. Alharthi, M.R. Karim, H. Misran, N. Amin, K.B. Sopian, Fabrication techniques and morphological analysis of perovskite absorber layer for high-efficiency perovskite solar cell: A review, Renew. Sus. Energy Rev. 98 (2018) 469-488. https://doi.org/10.1016/j.rser.2018.09.016

[20] P. Kajal, K. Ghosh, S. Powar, Manufacturing techniques of perovskite solar cells, in: H. Tyagi, A.K. Agarwal, P.R. Chakraborty, S. Powar (Eds.), Applications of Solar Energy, Springer, 2018, pp. 341-364. https://doi.org/10.1007/978-981-10-7206-2_16

[21] A. Kojima, K. Teshima, Y. Shirai, T. Miyasaka, Organometal halide perovskites as visible-light sensitizers for photovoltaic cells, ACS Pub. 131 (2009) 6050-6051. https://doi.org/10.1021/ja809598r

[22] D. Li, J. Shi, Y. Xu, Y. Luo, H. Wu, Q. Meng, Inorganic-organic halide perovskites for new photovoltaic technology, National Sci. Rev. 5 (2018) 559-576. https://doi.org/10.1093/nsr/nwx100

[23] D. Lin, T. Zhang, J. Wang, M. Long, F. Xie, J. Chen, B. Wu, T. Shi, K. Yan,W. Xie, Stable and scalable 3D-2D planar heterojunction perovskite solar cells via vapor deposition, Nano Energy. 59 (2019) 619-625. https://doi.org/10.1016/j.nanoen.2019.03.014

[24] L. Ma, W. Li, K. Yang, J. Bi, J. Feng, J. Zhang, Z. Yan, X. Zhou, C. Liu, Y. Ji, A-or X-site mixture on mechanical properties of APbX3 perovskite single crystals, APL Mater. 9 (2021) 041098-041112. https://doi.org/10.1063/5.0015569

[25] K. Mahmood, S. Sarwar, M.T. Mehran, Current status of electron transport layers in perovskite solar cells: Materials and properties, RSC Adv. 7 (2017) 17044-17062. https://doi.org/10.1039/C7RA00002B

[26] M. MGreen, E. Dunlop, D. Levi, J.H. Ebinger, M. Yoshita, A.H. Baillie, Solar cell efficiency tables (version 54), Prog. Photovolt. Res. Appl. 27 (2019) 565-575. https://doi.org/10.1002/pip.3171

[27] G. Niu, X. Guo, L. Wang, Review of recent progress in chemical stability of perovskite solar cells, J. Mater. Chem: A, 3 (2015) 8970-8980. https://doi.org/10.1039/C4TA04994B

[28] G. Niu, W. Li, F. Meng, L. Wang, H. Dong, Y. Qiu, Study on the stability of CH3NH3PbI3 films and the effect of post-modification by Aluminum oxide in all-solid-state hybrid solar cells, J. Mater. Chem: A, 2 (2014) 705-710. https://doi.org/10.1039/C3TA13606J

[29] J.H. Noh, S.H. Im, J.H. Heo, T.N. Mandal, S.I. Seok, Chemical management for colorful, efficient, and stable inorganic-organic hybrid nanostructured solar cells, Nano Lett. 13 (2013) 1764-1769. https://doi.org/10.1021/nl400349b

[30] M.F.M. Noh, C.H. Teh, R. Daik, E.L. Lim, C.C. Yap, M.A. Ibrahim, N.A. Ludin, A.R.M. Yusoff, J. Jang, M.A.M. Teridi, The architecture of the electron transport layer for a perovskite solar cell, J. Phys. Chem: C, 6 (2018) 682-712. https://doi.org/10.1039/C7TC04649A

[31] L.K. Ono, E.J.J.-Perez, Y. Qi, Progress on perovskite materials and solar cells with mixed cations and halide anions, ACS Appl. Mater. & Interfaces. 9 (2017) 30197-30246. https://doi.org/10.1021/acsami.7b06001

[32] N.-G. Park, M. Grätzel, T. Miyasaka, K. Zhu, K. Emery, Towards stable and commercially available perovskite solar cells, Nature Energy. 1 (2016) 1-8. https://doi.org/10.1038/nenergy.2016.152

[33] A. Poglitsch, D. Weber, Dynamic disorder in Methylammonium Trihalogenoplumbates (II) observed by millimeter-wave spectroscopy, AIP. 87 (1987) 6373-6378. https://doi.org/10.1063/1.453467

[34] A. Purabgola, B. Kandasubramanian, Thin films for planar solar cells of organic-inorganic perovskite composites, in: I. Khan, A. Khan, M. A. Khan, S. Khan, F. Verpoort, A. Umar (Eds.), Hybrid Perovskite Composite Materials, Elsevier, 2021, pp. 95-115. https://doi.org/10.1016/B978-0-12-819977-0.00003-2

[35] R. Sheng, A.H. Baillie, S. Huang, S. Chen, X. Wen, X. Hao, M.A. Green, Methylammonium Lead Bromide perovskite-based solar cells by vapor-assisted deposition, J. Phys. Chem. Lett. 119 (2015) 3545-3549. https://doi.org/10.1021/jp512936z

[36] Z. Shi, A.H. Jayatissa, Perovskites-based solar cells: A review of recent progress, materials and processing methods, Mater. 1 (2018) 700- 729. https://doi.org/10.3390/ma11050729

[37] H.J. Snaith, A. Abate, J.M. Ball, G.E. Eperon, T. Leijtens, N.K. Noel, S.D. Stranks, J.T.-W. Wang, K. Wojciechowski, W. Zhang, Anomalous hysteresis in perovskite solar cells, J. Phys. Chem. Lett: T, 5 (2014) 1511-1515. https://doi.org/10.1021/jz500113x

[38] W. Tress, N. Marinova, O. Inganäs, M.K. Nazeeruddin, S.M. Zakeeruddin, M. Graetzel, The role of the hole-transport layer in perovskite solar cells - Reducing recombination and increasing absorption in 2014 IEEE 40th Photovoltaic Specialist Conference (PVSC), IEEE. (2014) pp. 1563-1566. https://doi.org/10.1109/PVSC.2014.6925216

[39] W. Tress, N. Marinova, T. Moehl, S.M. Zakeeruddin, M.K. Nazeeruddin, M. Grätzel, Understanding the rate-dependent J-V hysteresis, slow time component, and aging in CH3NH3PbI3 perovskite solar cells: The role of a compensated electric field, Energy & Environ. Sci. 8 (2015) 995-1004. https://doi.org/10.1039/C4EE03664F

[40] E.L. Unger, A.R. Bowring, C.J. Tassone, V.L. Pool, A. G.-Parker, R. Cheacharoen, K.H. Stone, E.T. Hoke, M.F. Toney, M.D. McGehee, Chloride in Lead Chloride-derived organo-metal halides for perovskite-absorber solar cells, ACS Publ. 26 (2014) 7158-7165. https://doi.org/10.1021/cm503828b

[41] H. Wang, M. Zhou, P. Choudhury, H. Luo, Perovskite oxides as bifunctional oxygen electrocatalysts for oxygen evolution/reduction reactions: A mini review, Appl. Mater. Today. 16 (2019) 56-71. https://doi.org/10.1016/j.apmt.2019.05.004

[42] Q. Wang, N. Phung, D.D. Girolamo, P. Vivo, A. Abate, Enhancement in lifespan of halide perovskite solar cells, Energy & Environ. Sci. 12 (2019) 865-886. https://doi.org/10.1039/C8EE02852D

[43] J. Wei, Y. Zhao, H. Li, G. Li, J. Pan, D. Xu, Q. Zhao, D. Yu, Hysteresis analysis based on the ferroelectric effect in hybrid perovskite solar cells, J. Phy. Chem. Lett. 5 (2014) 3937-3945. https://doi.org/10.1021/jz502111u

[44] W.S. Yang, J.H. Noh, N.J. Jeon, Y.C. Kim, S. Ryu, J. Seo, S.I. Seok, High-performance photovoltaic perovskite layers fabricated through intramolecular exchange, Science. 348 (2015) 234-1237. https://doi.org/10.1126/science.aaa9272

[45] X. Zhao, N.-G. Park. Stability issues on perovskite solar cells, Photonics. 2 (2015) 1139-1151https://doi.org/10.3390/photonics2041139

Perovskite based Materials for Energy Storage Devices   Materials Research Forum LLC
Materials Research Foundations 151 (2023) 1-32   https://doi.org/10.21741/9781644902738-1

[46] D. Zhou, T. Zhou, Y. Tian, X. Zhu, Y. Tu, Perovskite-based solar cells: Materials, methods, and future perspectives, J. Nanomaterials. 8148072 (2018) 1-5https://doi.org/10.1155/2018/8148072

Chapter 2

Organometallic Halides-Based Perovskite Solar Cells

Uzma Hira[1*] and Muhammad Husnain[1]

[1]School of Physical Sciences (SPS), University of the Punjab, New Campus, 54590, Lahore, Pakistan

uzma.sps@pu.edu.pk

Abstract

The energy crisis is increasing day by day and the natural resources of energy are rapidly decreasing from all around the world. The utilization of solar energy has become popular in the last few years. But the silicone-based solar cells (SCs) are too costly and not easily bought by everyone. A lot of experiments have been performed to make less expensive and more stable solar cells. Currently, scientists are trying to develop SCs through the utilization of organic and inorganic ionic materials. These compounds are usually considered organometallic halides-based perovskite solar cells (OMHP-SCs). They have properties of semiconductors and a remarkable *p-n* junction. Perovskite solar cells (PSCs) have gained the attention of researchers owing to their excellent power conversion efficiency (PCE) and the maximum efficiency obtained so far is ~ 30%. But their instability in the environment ceases the practical application of OMHP-SCs. Therefore, in order to overcome the degradation of PSCs, a large number of passivation methods have been proposed such as suppression of ions, solvent engineering, hole conductor-free perovskite, etc. It is anticipated that OMHP-SCs will be available in the market in the coming years with great stability and PCE.

Keywords

Solar Energy, Perovskite Solar Cells, Organometallic Halide-Based Perovskite Solar Cells, Power Conversion Efficiency, Passivation Methods

Contents

Organometallic Halides-Based Perovskite Solar Cells..................................33

1.   Introduction..35

      1.1   Carbon-based energy sources 36

1.2   The global trend toward renewable energy resources 36

1.3   Era of Solar Cell (SCs) technology 36

1.4   Green energy (Carbon free) 37

2.   Photovoltaic effect 37

2.1   Discovery of Sir Alexander Edmond Becquerel 38

2.2   Development of solar cells 38

2.3   Generations 39

2.4   Types of 3rd generation of SCs 40

3.   Perovskite-based solar cells 40

3.1   Introduction to perovskite compounds 40

3.2   Classification of perovskite 41

3.3   Organometallic halide-based perovskite (OMHP) solar cells 41

3.4   Evolutionary history of perovskite solar cells with their
efficiency 42

3.4.1   Open-circuit voltage (OCV) 44

3.4.2   Short-circuit voltage (J_{sc}) 44

3.4.3   Fill factor (FF) 44

3.5   Crystal structure of organometallic halides-based perovskite
solar cells 44

3.6   Behavior of OMHP with different combinations of A, B,
and X 45

3.6.1 A-site cations 45

3.6.2 B-site cations 45

3.6.3 X-site anions 46

3.6.3.1 Iodide (I) anion 46

3.6.3.2 Chloride (Cl) anion 46

3.6.3.3 Bromide (Br) anion 46

3.7   Goldschmidt tolerance factor (τ) 46

3.8   Octahedral factor (OF) 47

**4.   Important Parameters of Organometallic Halide-Based Perovskite
(OMHP) 47**

4.1   Charge transport (CT) 47

4.2   Diffusion length and mobility of charge carriers 48

4.3   Electronic structure (ES) 49

4.4   Effect of effective masses of holes and electron carriers 49

5.   Environmental instability of organometallic halides-based perovskites (OMHPs) solar cells 50

5.1   Degradation and stability issue...50

5.2   Effect of moisture 50

5.3   Effect of temperature 51

5.4   Effect of oxygen and light 52

6.   Recent development in the OMHP solar cells...................................53

6.1   Ion migration and the suppression of ions......................................53

6.2   Solvent engineering 54

6.3   Annealing...54

6.4   2D/3D technology...54

6.5   Organometallic halides-based perovskite quantum dot solar cells55

6.6   Solid-state hole conductor-free (HCF) OMHP-SCs.........................57

6.7   Tandem perovskite solar cells (TPSCs) 58

6.8   Passivation of OMHP-SCs 58

Conclusion...59

References..59

1.   Introduction

The use of energy is rising throughout the world and the need for energy will reach its highest point in the coming years. Particularly for developing countries, the utilization of energy has become a critical issue, where consumption of energy is greater than production. Researchers are trying to solve this problem. They are trying to develop affordable, renewable, non-polluting, as well as free from toxic oxides of nitrogen forms of carbon energy sources [1]. From a sustainable energy origin, the use of solar energy as electrical energy is an interesting part of the research. Solar energy (SE) is the most easily available all around the map, an unlimited and eco-friendly source for human beings. Scholars are making an effort to find an economically affordable and sustainable procedure to convert photons of sunlight into electrical signals [2].

Utilization of energy without using carbon resources is the need time to save the resources of fossil fuel. Carbon-free energy has also a great effect on our green environment. The movement towards solar energy is the best way to conserve the reserves of carbon as well as fulfill the requirements of energy in developing countries. For this purpose, we must have to commercialize solar cell technology [3,4].

1.1 Carbon-based energy sources

Currently, one of the greatest threats to the world is deteriorating environmental quality. The CO_2 emission assesses the environmental hazards. The reason is that CO_2 is an important source of GHGs (greenhouse gases). CO_2 is not always a suitable indicator for environmental hazards [5]. The main contributor to CO_2 gas is the energy sector. The international organization "Intergovernmental Panel on Climate Change" (IPCC) reported that the temperature of the Earth is continuously increasing due to the rapid growth of oxides of carbon and nitrogen emissions in Mother Nature. The one hundred and ninety-six states signed in the Paris meeting to control the pollution and temperature of the planet. They agreed to take action to control the earth's surface temperature from pre-industrial levels to below 2°C [6]. Solar cell technology is an emerging and the best way towards a sustainable source of energy, to overcome global warming and currently prevailing energy shortfall issues due to the depletion of fossil fuels.

1.2 The global trend toward renewable energy resources

Energy shortage is one of the most critical issues of the 21st century. In the past, researchers had been trying to find an alternate way to satisfy global energy needs but still, they are unable to commercialize such a sustainable setup that fulfills the requirements of energy [7]. According to the "International Energy Outlook" (IEO), the utilization of marketed energy by the total world is projected to increase by 44% within 25 years. International companies produce energy from renewable energy means. The utilization of non-fossil fuel energy is one of the main concerns of every country in the world because the use of carbon in the world is increasing day by day which is dangerous for our green environment. The resources of carbon are also decreasing rapidly all around the world. If we are unable to control the use of carbon, we will have to face energy problems shortly [8].

1.3 Era of Solar Cell (SCs) technology

When there was no source of energy in the world, the sun was the only source of energy. It is not wrong to say that all types of energy on the earth are directly connected with the sun. Solar energy (SE) is available all over the world. Electrical energy is also generated from the sun with the help of different techniques. In this chapter, our main concern is how we can get and increase the efficiency of solar cells to get a high-voltage electric signal.

Sunlight is the only source from all other renewable energy resources which is converted into electrical energy. Nearly 1.8×10^{11} MW, energy captured by the earth is a huge amount of energy and larger than all other sources. A lot of solar-based technologies are developed to convert SE into electrical energy. SCs can be used in cooking, refrigeration, thermoelectric power station, dying, and photovoltaic devices [9].

1.4   Green energy (Carbon free)

A well-known person said:

"You are thus independent of fossil fuels and at the same time make an active contribution to climate protection" [10].

Due to the increasing rate of global warming, it is necessary to create new strategies for clean energy to control our green environment. The emission of greenhouse gases such as CO_x, NO_x, etc. promotes global warming. We have to commercialize such types of systems which control the emission of dangerous gases and generate non-polluted energy sources such as sunlight, wind, rain, waves, etc. [11]. After the COVID-19 pandemic period, every state is trying to develop a clean and green environment and thus struggling to reduce pollution sources. During the lockdown, the pandemic peak period has left certain effects on society and the economy. It has helped to repair some environmental problems. The percentage of greenhouse gases (GHGs), oxides of nitrogen (NO_x) and carbon (CO_x), polluted water, noise pollution, etc. have decreased appreciably due to complete or partial lockdowns and limited activities all around the world [12].

SE has several benefits as it is non-polluting, renewable, low cost, and inexhaustible. But with advantages, this energy has also some disadvantages that include variability, instability, low power, high technology requirements, etc. Due to these reasons, photovoltaic energy is not commercialized on a large scale. Due to the rapid improvement in photovoltaic technologies and their accelerated cost reduction, they are becoming an interesting part of developing countries [13,14]. There are a few limitations with this technology like the unavailability of solar energy during nighttime and its non-uniformity all over the world.

2.   Photovoltaic effect

The Photovoltaic (PV) effect is the backbone of solar cell discovery. It may be defined as the transformation of a photon's energy into an electrical signal (ES) called the photovoltaic effect. PV is an interesting scientific phenomenon for the change of SE directly into ES.

PV is a device composed of p-n junctions that have the properties of semiconductors. This system converts SE into electricity based on the principle of the PV system as shown in Fig. 1 [15]. When a PV cell is exposed to light, photons are saturated by the crystal of conducting materials and remove the substantial free electrons in the crystal that generate the electricity [16]. Therefore, PV has a lot of applications in diverse fields where energy is needed in small volts like road lights, water electric pumping, electric vehicle batteries, satellites, and other electric vehicles.

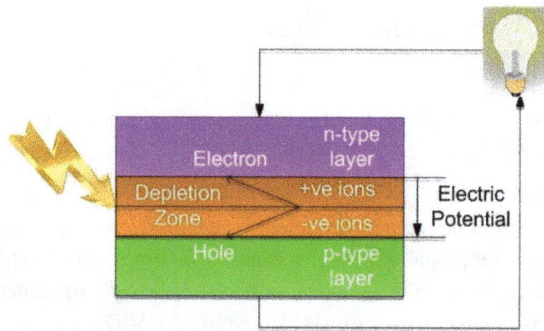

Figure 1. *This figure shows the phenomenon of conversion of light into electricity. It is a PV device having p-n junction, this junction is basically captured and convert its energy into electricity (Reproduce with copyright permission) [15].*

2.1 Discovery of Sir Alexander Edmond Becquerel

Alexander Edmond Becquerel (French researcher) in 1839, discovered and first time developed a cell to convert the photon's energy into electricity which is known as the PV effect. In the last few years, a marvelous advancement has been developed in PV technology in order to improve the efficiency of PV cells [15].

2.2 Development of solar cells

In 1954, Bell Labs proclaimed the silicon SCs and in 1957, Bell with Pearson, Chapin, and Fuller got a patent to develop silicon-based SC (with an 8% efficiency rate). This work was now led to the advanced development of a diverse variety of PV cells. Usually, it can be defined as the surfacing of an electric voltage between two electrodes attached to a solid or liquid system. It is basically a p-n junction type cell as shown in Fig. 2. Practically, all photovoltaic cells incorporate a p-n junction in a semiconductor across which the photovoltage is applied [17].

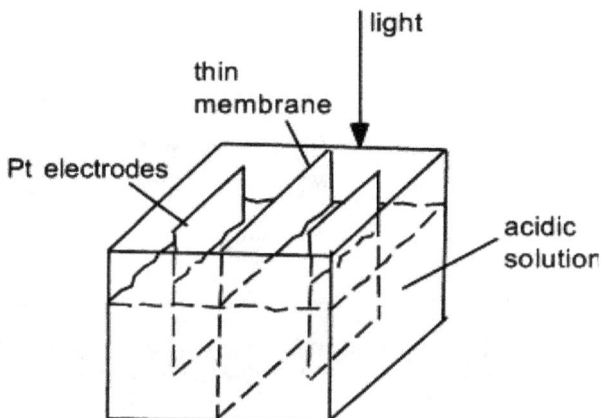

Figure 2. *This diagram represented apparatus used by Becquerel (Reproduce with copyright permission) [17].*

2.3 Generations

The composition of solar cells is still a challenging task for researchers and development companies for SCs. In this chapter, we are going to discuss especially organometallic halide-based perovskite Solar Cells (OMHP-SCs). For the composition of SCs, a lot of experiments have been done and different materials are suggested. Based on different types of materials and their composition, three generations of solar cells have been developed with different power conversion efficiencies (PCE). The hierarchy of development of SCs is classified as first, second, and third generations. OMHP-SCs are regarded as a third generation of SCs. They have intrinsic properties to convert SE into ES and their PCE reaches up to ~ 30%. Scientists are interested in this type of SCs due to their material availability and economic viability as compared to first and second generations. Nowadays the third generation based SCs are being prepared. However, sun energy converted into electricity through photovoltaic technology is still a main task for all research groups and development sectors due to instability. Table 1 is showing the basic differences between different types of SCs generations [18,19].

Table 1. *Basic difference between the 1ˢᵗ, 2ⁿᵈ And 3ʳᵈ generation of solar cells (Self-Drawn).*

Characteristics	Generation-1	Generation-2	Generation-3
Composition	Silicon-wafers	Silicon-thin film	Nanomaterial-Based
Conductivity type	Photovoltaic	Photovoltaic	Photovoltaic
Material availability	Not easily available	Not easily available	Easily available
Stability	More stable	More stable	Less stable
Examples	Poly-crystalline Si wafers	Thin silicon films on indium tin oxide etc.	Quantum Dots, Perovskites etc.

2.4 Types of 3ʳᵈ generation of SCs

1. Organic SCs

2. Dye-sensitized SCs

3. Nanocrystal SCs

4. Polymer SCs

5. Perovskite SCs

In this chapter, we will discuss only the perovskite solar cells (organometallic halides-based perovskite). A few organizations are involved to develop perovskite-based SCs with great efficiency. Nowadays PCE rises to 30 % or more [20-22]. In the following section, a detailed description of perovskite-based SCs is provided.

3. Perovskite-based solar cells

3.1 Introduction to perovskite compounds

- Calcium titanium dioxide ($CaTiO_2$), in 1839, was discovered by scientist Gustav Rose belonging to the state of Russia. It is a mineral compound and the research on compounds of this group was continued by another Russian mineralogist Lev Perovski, that's why these are called Perovskite compounds. Later on, all those compounds that had a crystal structure similar to $CaTiO_3$ were categorized as perovskite materials.

- Material research departments are interested in perovskite due to its huge availability in nature and low-cost fabrication as well.

- At present, a large number of compounds are known that adopt the perovskite structure.

- The most abundant solid perovskite in the earth's interior is Bridgmanite (Fe, Mg)SiO$_3$, making up 38% of the total [23].

3.2 Classification of perovskite

Usually, perovskite materials are represented by formula ABX_3, where A & B are cations and atoms of B-site are smaller than atoms of A-site and X is the anionic site. Based on the X group, perovskites are classified into different categories as shown in Fig. 3.

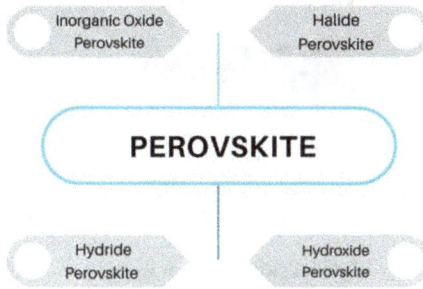

Figure 3. *This classification is based on the X (halogen) site of perovskite (Self-Drawn).*

3.3 Organometallic halide-based perovskite (OMHP) solar cells

From the last few years, organometallic halides-based perovskites SCs have become the great priority of researchers due to their markedly PCE. OMHP-SCs are regarded as high-efficiency photovoltaic cells [24]. They exhibit semiconducting-like behavior, high extinction coefficient, and broad light absorption range. Therefore, OMHP materials are considered an interesting/robust candidate for photo absorption. They are suitable for the photovoltaic effect because their band gap is reversely related to their dimensions.

The first report on photovoltaic perovskite cells was provided by Miyasaka and co-workers [25]. They have reported CH$_3$NH$_3$PbBr$_3$ PV cells with an efficiency of 2.2 %. They were attracted by the self-organization potential of perovskites in the nonporous TiO$_2$ layer of dye-sensitized cells. Characterization of PSCs compounds can be enhanced by replacing cations and anions at the ABX_3 sites of the perovskite with other compounds. The general structure of *ABX₃* type perovskite is shown in Fig. 4. In the literature so far different OMHP materials have been studied for SCs applications such as $CH_3NH_3PbI_3$, $CH_3NH_3PbI_{3-x}Cl_x$, $CH_3NH_3PbBr_3$, CH_3NH_3Pb $(I_{1-x}Br_x)_3$, $HC(NH_2)_2PbI_3$, $CH(NH_2)_2$ $Pb(I_{1-x}Br_x)_3$, $CH_3NH_3SnI_3$ materials [26].

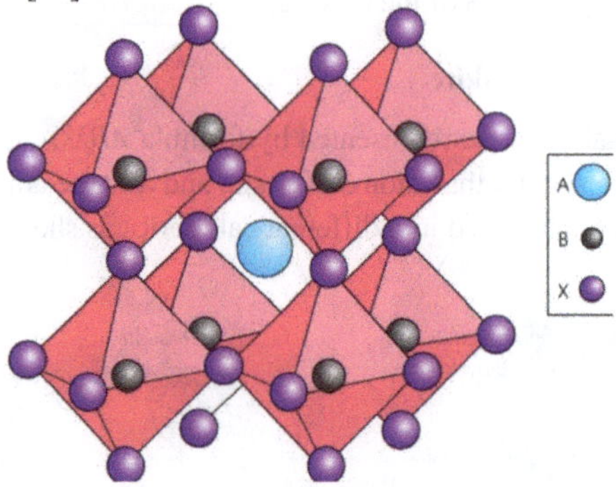

Figure 4. *Diagram represents the simple crystal structure of ABX₃ type perovskites (Reproduce with copyright permission) [26].*

3.4 Evolutionary history of perovskite solar cells with their efficiency

Over the last two decades, the trend of non-fossil fuels/renewable energy sources is increasing rapidly. Scientists have been trying to find a non-polluted, cheap, sustainable, and easily available material for energy production. SCs development is the most emerging trend toward sustainable sources of energy. For the development of SCs, material availability or its property to convert sunlight into electricity at low cost is one of the most important concerns of scientists. Different types of SCs have been proposed with different power conversion efficiencies. Silicon-based SCs are best on account of PCE but their availability is too costly. OMHPs are also proposed for photovoltaic SCs but at initial levels, their PCE is too low now with different passivation techniques their PCE has increased up to 25-30 % [27].

Therefore, SC is the most efficient and convenient method to utilize SE as a source of electricity. The maximum capacity of SCs for the conversion of photon energy into electrical energy is regarded as their power conversion efficiency. The PCEs of some important OMHP are listed in Table 2 [21].

Table 2. *Power conversion efficiency of different organometallic halides-based perovskite by using different protective layers and with device fabrication (Reproduce with copyright permission) [21].*

Power conversion efficiency (PCE) / %	V_{oc} /v	J_{sc} / $mAcm^{-2}$	FF	Module Conformation	Time period
3.8	0.61	11	0.57	Platinum-FTO/Electrolytic solutions/$CH_3NH_3PbI_3$/TiO_2	(2009)
6.54	0.706	15.82	0.586	Platinum/fluid electrolytic solutions $CH_3NH_3PbI_3$ (QD) / TiO_2/FTO	(2011)
9.7	0.888	17	0.62	Platinum-FTO/Electrolytic solutions/ methyl ammonium lead iodide/TiO_2	(2012)
15	0.993	20	0.73	Au/spiro OMeTAD/ methyl ammonium lead iodide /mTiO_2/FTO	(2013)
17.01	1.056	21.64	0.741	Gold/spiroMeOTAD/TiO_2/$CH_3NH_3PbI_3$/Glass	(2014)
19.3	1.114	23	0.74	Au/spiro-MeOTAD/Cubic methyl ammonium lead iodide (MAPbI$_3$)/ MAPbI$_3$/m-TiO_2/c-TiO_2/FTO	(2014)
20.2	1.06	24.7	0.775	Gold/Spiro-OMeTAD/Perovskites/m-Li:TiO_2/ Perovskites/ FTO	(2015)
20.5	1.114 3	23.24	0.759	Gold/PTAA/Perovskites/(bl/m-TiO_2)/FTO	(2016)
21.6	1.14	23.7	0.78	Gold/Spiro-OMeTAD/Perovskites/m-TiO_2/ bl-TiO_2/ FTO	(2016)
22.5	1.11	25	0.817	Gold/Spiro-OMeTAD/Perovskites/m-TiO_2/ Perovskites/bl-TiO_2/ FTO	(2018)
22.7	1.14	24.92	0.792	Perovskites combined sheet/perovskites upper sheet /PTAA/gold/ FTO/TiO_2/m-TiO_2	(2019)

The PCE value of these types of cells has rapidly increased, within a short time range the value increased from 3.8 to 28.2%. Research on OMHP-based perovskites is continuing and researchers are trying to commercialize them on a large scale [28]. To calculate the

PCE of a PSC, generally, we use the formula ($PCE = \times J_{sc}\ V_{oc} \times FF$) [28]. Where V_{oc} is used for open-circuit voltage, J_{sc} represents the short-circuit voltage and FF stands for fill factor. These parameters are described in detail below:

3.4.1 Open-circuit voltage (OCV)

The voltage created by the open circuit terminals of a PV cell is named its open circuit voltage (OCV) and is represented by the symbol V_{oc}. The value of V_{oc} depends on the procedure of fabrication of PV cells and temperature. The intensity of incident light and the surface area of the cell available for the sun have no significant effect on the value of V_{oc}. Frequently, the numerical value of V_{oc} of PV devices ranges from 0.5 to 0.6 V [29].

3.4.2 Short-circuit voltage (J_{sc})

The J_{sc} value is the light absorption and photocurrent conversion capacity of SCs. Its value fluctuates with the thickness of perovskite material as well as is affected by the incomplete film coverage owing to decreased absorption [30].

3.4.3 Fill factor (FF)

To measure the quality of SCs, the fill factor (FF) term is used. It can be determined by comparing the maximum power with the theoretical power that would be the output at both the J_{sc} and V_{oc} together.

Presently, the best experimental values of J_{sc}, V_{oc}, and FF recorded in PSCs are 27.4 mAcm^{-2}, 1.5 V, and 86.7 % respectively. These are too high values and cannot be recorded simultaneously in a single PSC.

3.5   Crystal structure of organometallic halides-based perovskite solar cells

ABX_3 is a general representation of organometallic halides perovskites, where:

- A-site cation is larger than B-site cation, B-site cation occupies a center of cubic-octahedral and A-site cation is surrounded by neighboring (12 coordination number) anions.

- B cation is small in size and shows greater stability in an octahedral shape. Its site is shared with 6-X anions. Cation B has six nearest neighbors (VI coordination number).

- X belongs to the halogen group (F, Cl, Br, I).

3.6   Behavior of OMHP with different combinations of A, B, and X

3.6.1 A-site cations

The electronic properties of the perovskite material can be changed by changing the cation of the A-site. The crystallography of crystals of perovskites is also changed by the size of the A-site cation [31,32]. By replacing the smaller cation with the larger one the crystal symmetry is ultimately changed and gives the low band gap and good light absorption [33]. By changing the 'A' site cation with small cations of organic molecules like

- Methyl ammonium (MA)
- Ethyl ammonium (EA)
- Formamidinium (FA)
- Guanidinium
- Dimethyl ammonium
- Hydrazinium
- Hydroxyl ammonium

Sometimes mixed cations can be used such as MA & EA. The most frequently used cation in OMHP material is methyl ammonium (MA).

3.6.2 B-site cations

Mostly, the group IV-A elements of the periodic table are used in the B-site of the crystal in OMH-perovskite. Usually, the oxidation state (OS) of B type metal is +2. For instance, lead (Pb), tin (Sn), and germanium (Ge) are common examples. From all of the elements of group IV-A, lead metal is frequently used and such compounds are regarded as lead halide perovskite SCs. The compounds having lead metal show excellent PV effect and stability as compared to the tin metal (Sn). But one problem associated with lead metal is that due to an increase in ionic characters moving from top to bottom in groups in the periodic table, the band gap is decreased. Hypothetically, the band-gap values of $MASnX_3$ and $MAPbX_3$ are 1.2–1.4 and 1.6–1.8 eV respectively but $MASnX_3$ is unstable in the +2 oxidation state and less attractive metal for OMHP SCs. Therefore, oxidation of PSC in the air is easy. They are changed into volatile SnI_4 compounds having +4 OS. $MASnX_3$-based PSCs exhibited better value J_{sc} as compared to the lead-based SCs like $MAPbX_3$ but were not commercialized due to their environmental instability [31].

3.6.3 X-site anions

Different halide ions are the best option to change the characteristics of OMH-perovskite solar cells. The absorption spectra for halogen ions shifted towards longer wavelengths and a decrease in energy is observed from top to bottom in the seventh group of the periodic table. The behavior of each anion of group VIIA has been explained below:

3.6.3.1 Iodide (I) anion

The OMH-perovskite SCs based on iodine anions show the PCE over 20%. In most cases we frequently use iodine, the reason being the electronegativity difference. The values for iodine and lead are 2.66 and 2.33, respectively. Due to this difference in electronegativity, iodine-based OMHP solar cells are the most stable perovskite structures.

3.6.3.2 Chloride (Cl) anion

The OMHP with chloride anion perovskite crystal structure exhibits longer carrier diffusion lengths and better charge carrier lifetimes with good PCE. For instance, at room temperature, the crystal structure of $MAPbCl_3$ is cubic. With the help of the combination of Cl and I in crystal $CH_3NH_3PbI_{3-x}Cl_x$, has been an excellent aligned crystalline arrangement with greater PCE.

3.6.3.3 Bromide (Br) anion

The band gap of the perovskite SCs is effectively tuned by using Br at the X-site of ABX_3. With the introduction of bromide anions with a combination of iodine in OMHP, the band gap is increased as a result of the crystal being distorted. For example, when only Br is used like $MAPbBr_3$ crystal structure is cubic at room temperature but with iodine combination like $CH_3NH_3PbI_{3-x}Br_x$ compound showed excellent PCE, and stability. By changing the A-site, MA with FA ($FAPbI_{3-x}Br_x$) inconsistent band gap was created. With a range of film colors, its value ranges from 1.48 to 2.23 eV [32].

3.7   Goldschmidt tolerance factor (τ)

Some early investigations on perovskite materials were carried out by Victor Moritz Goldschmidt in 1926. Goldschmidt's tolerance factor (τ) provides information about the stability and distortion of perovskite crystal structure. τ also gives the relationship between cationic and anionic sizes of A, B, and X in ABX_3 type compound with an empirical relationship which can be expressed by using the relation given below:

$$\tau = \frac{r_A + r_X}{\sqrt{2}(r_B + r_X)} \tag{1}$$

Perovskite based Materials for Energy Storage Devices   Materials Research Forum LLC
Materials Research Foundations 151 (2023) 33-66   https://doi.org/10.21741/9781644902738-2

The ionic radii of the A, B, and X-sites of ABX_3 are represented by r_A, r_B, and r_X respectively, in the crystals of PSCs. The tolerance factor (τ) ranges from about 0.77 to about 1.06 and its value is 1 for a perfectly ideal cubic perovskite crystal structure. In the development of new PV devices for practical applications, the stability of OMHP is a challenging path for researchers. With the help of the Goldschmidt tolerance factor, the stability of the OMHP-SCs is predicted [33]. However, Goldschmidt's tolerance factor does not explain entirely and there are too many exceptions [34].

3.8 Octahedral factor (OF)

The octahedral factor (OF) is also used to predict the stability of crystal structures of OMHP-SCs. The OF can be defined as the ratio radii of cation B to halide group X as shown below in eq. 2:

$$\mu = \frac{r_B}{r_X} \tag{2}$$

Normally, when the value of μ is greater than 0.4 ($\mu > 0.41$) the crystal structure of OMHP-based SC is stable. The OF is also called the constraint factor [35].

4. Important Parameters of Organometallic Halide-Based Perovskite (OMHP)

To develop a new generation of PV cells with higher PCE, a diverse variety of materials have been suggested to develop SCs but the scientists are unsuccessful to commercialize them due to the fewer PCEs as compared to the silicon-based PV SCs. The 3rd generation SCs with higher PCEs have been developed during the last 25 years. Recently, OMHPs have become extremely attractive materials in the development of SCs with low-cost manufacturing. They exhibit reasonable performance. Due to their low cost, non-polluting process, and availability, companies and different researchers are still trying to commercialize them [36].

4.1 Charge transport (CT)

The OMHPs show well-balanced electron and hole transport mechanisms. By using different techniques, the charge transport (CT) properties of OMHPs can be enhanced. The CT mechanism is an important criterion in the development of high PCE of OMHP SCs [42]. OMH-perovskites show both n and p-type junction properties. CT is still challenging for thin-film PSCs. Photovoltaic systems in OMHP act as an absorber in the SC configuration. They can provide an efficient transport mechanism for both electrons and holes. The high rate of PCE of OMHPs is also due to their good CT mechanism for electrons as well as holes [37]. The CT properties of OMHP mainly depend upon the:-

- Electronic Structure
- Diffusion length (DL)
- Mobility of charge
- Combination of different halides
- Effective masses ratio of electron and hole

Electronic clouds of the material especially in semiconductors type materials directly influence the CT mechanism. Importantly, the effective magnitude of the electron and hole is inversely proportional to the curve of the electronic band structure. Usually, methylammonium lead triiodide perovskite (MAPbI$_3$) or the combination of iodine and bromine is used in OMHP solar cells. OMHPs are characterized by the ambipolar behavior of electron and hole transport mechanisms. With the help of different analytical techniques, like femtosecond transient optical spectroscopy, the researchers can find the CT mechanism of different OMHPs. For instance, the electron-hole DL is 100 nm for the solution-processed MAPbI$_3$. This value can be compared with the optical absorption length. Long electron-hole DL indicates that more electrons and holes can reach PSCs electrodes to develop an external electrical signal. OMHPs-SCs have unique behavior and they show ultra-long DL [38].

4.2   Diffusion length and mobility of charge carriers

The value of diffusion length (DL) is extremely important for both single as well as mixed halide perovskites SCs. Its value can be varied by changing the composition of the film, crystal structure, and grain size [39]. For instance, in the starting era of OMHPs, it was investigated that the combination of two halides in OMHPs-SCs like MAPbI$_{3-x}$Cl$_x$ had a longer DL (DL > 1 μm). The DL values for mixed and single OMHP are > 1 and ~ 100 nm, respectively [40]. It is always not necessary that the hole and electron DLs are balanced as in the CH$_3$NH$_3$PbI$_3$ compound, where hole DL is 813 ± 30 nm and electron DL is 813 ± 30 nm. When the compound is prepared by a sequential deposition method, the hole DL has a greater value than the electron DL in CH$_3$NH$_3$PbI$_3$. The mobility and lifetime of the charge carriers in SCs are determined directly with the help of the DL of any PSC. The diffusion length of carriers can be given by using the following relationship

$$DL = D\tau_e \qquad (3)$$

Where D is the diffusion coefficient and τ_e is the lifetime of charge carriers. Further diffusion coefficient can be represented by considering carrier's mobility using the formula:

$$D = \frac{\mu_q}{k_B T} \tag{4}$$

Where, μ_q = mobility of charge carrier k_B = Boltzman constant and T = temperature. The mobility values of charges are greatly influenced by the nature of compounds used for making SCs. Values from Hall-effect measurements are usually high ranging from 0.5 to 0 cm^{2}V^{-1}s^{-1}. The electron-DL for MAPbI$_3$ was 130 nm, and the hole-DL was 100 nm when a mixed halide combination was used like MAPbI$_{3-x}$Cl$_x$ then the estimated electronic DL was $\sim$ 1069 nm, whereas the DL of hole carriers was $\sim$ 1213 nm [41]. The ratio of effective masses of electrons and holes also affects the DL. For example, photo-generated electrons and holes in MAPbI$_3$ calculated by DFT are m$_o$ = 0.23 mh and m$_o$ = 0.29 mh, respectively [42].

4.3 Electronic structure (ES)

Band gaps and electronic structure play a vital role in PV technology. The efficiency of OMHP SCs is greatly varied by the band gap of the crystal constituents. DFT is a favorable tool for the determination of the ES of the crystal of PSCs [43]. To understand the mechanism of the PV system and the working of SCs, it is extremely important to find the band gap and the ES of crystals [50]. The ES, density of states (DOS), and the band gap of CH$_3$NH$_3$PbI$_3$ are determined with the help of DFT. From DFT, it was found that the electron of the 6s orbital of lead (Pb) was below the top of the valence band of PSCs. Therefore, it is said that the 6s electrons of the lead element show unexpected behavior and give unusual properties to OMHPs [44]. The ionic and covalent nature of OMHPs-SCs is due to the strong s and p antibonding character of lead and iodine respectively in VBM (valance band maximum) and the CBM (conduction band minimum) which is constituted almost completely from the p-orbitals of a lead atom. So, the CBM and VBM give unique characteristics to PSCs. The size of the cation in ABX$_3$ type crystal also affects the band gap as well as the ES of the perovskites SCs. The selection of the halide ion has a great effect on the valence band position and the p-orbitals change from 5p, 4p to 3p for I, Br, and Cl, respectively. Perovskite-solar cells behave as an efficient PV device because of their ES and tunable bandgap [45,46].

4.4 Effect of effective masses of holes and electron carriers

The effective mass ratio of carriers (electron and hole) is an important parameter for the development of highly efficient SCs. The ratio of the effective mass is the fundamental study of any semiconductor and this ratio is very helpful in determining the PCE of solar cells. Device performance is totally dependent upon the effective masses of carriers; a small tuning in this ratio conventionally deteriorates the quality of the crystal [46]. Effective masses basically represent the number or quantity of electrons and holes in semiconductors. The effective mass of electrons (m_e) and holes (m_h) represent the mobility of electrons (μ_e) and holes (μ_h) respectively. The value of m_e and m_h for MAPbI$_3$ is 0.32m$_o$ and 0.36m$_o$, respectively. Tetrahedral semiconductors exhibit the trend of effective masses as $m_e < m_h$ and ultimately $\mu_e > \mu_h$. particularly, the value of μ_e is significantly higher than μ_h where the difference between m_e is less than m_h. Interestingly, the value of effective masses for MAPbI$_3$ is nearly equal to the same ($m_e = m_h$) therefore, the value of charge mobility is also the same ($\mu_e = \mu_h$). In OMHP-SCs, the quite balanced ($\mu_e = \mu_h$) mechanism in MAPbI$_3$ provides a great advantage for efficient carrier collection [42,47].

5.   Environmental instability of organometallic halides-based perovskites (OMHPs) solar cells

The instability is the major challenging problem of PSCs technology. If we are able to form environmentally stable PSCs, it will be very fruitful to solve the problems of the energy crisis and protect our environment from polluting energy resources. Here, some methods are discussed to find a solution to the instability of OMHPs-SCs.

5.1   Degradation and stability issue

No doubt, OMHPs-SCs is a brand-new concept of PV technology with high PCE. But their environmental long-term stability under harsh conditions is still challenging and the major barrier to their commercialization. In the current situation, the excellent achievement of OMHPs-SCs in the research areas is because of low-cost production as well as the availability of material but SC stability has still been a serious issue in their practical application. Their lifetime is too short. If scientists are able to make stable perovskites that are able to bear harmful conditions of the environment like moisture, light, and temperature then they can commercialize these solar cells for use. By solving this problem, our energy reservoir might be saved and non-polluted energy can be produced [48]. Here, we discuss the main factors that make the OMHPs-SCs unstable.

5.2   Effect of moisture

Moisture has been considered the most challenging part of the stability of perovskite. Assorted mechanisms have been recorded on the degradation of OMHPs brought about by moisture. Formally, when the degradation process is started, the reaction proceeds in the forward direction until the equilibrium is established. At the equilibrium stage, the molecules of H_2O are in equilibrium with HI molecules and the vapor pressure of CH_3NH_2 reaches equilibrium as shown in Eq. 5 [49].

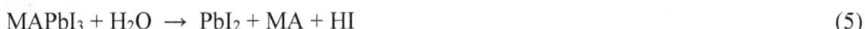

$$MAPbI_3 + H_2O \rightarrow PbI_2 + MA + HI \tag{5}$$

From the given equation, we can say that in the presence of H_2O, firstly, $MAPbI_3$ decomposes into MAI and PbI_2. The mechanism of degradation of $MAPbI_3$ in the moisture is shown in Fig. 5 [50]. The reaction does not stop until the MAI solution decomposes into CH_3NH_2 and HI. After the formation of HI, a photochemical reaction takes place in which HI reacts with oxygen and generates I_2. This reaction proceeds under the light.

Kamat and his coworkers [51] reported another concept for the degradation of PSCs. They proposed that $MAPbI_3$ can form a stable hydrated complex with H_2O like this $MA_4PbI_6 \cdot 2H_2O$ as shown in Eq. 6, rather than simply converting into PbI_2 under a moisturizing environment [51].

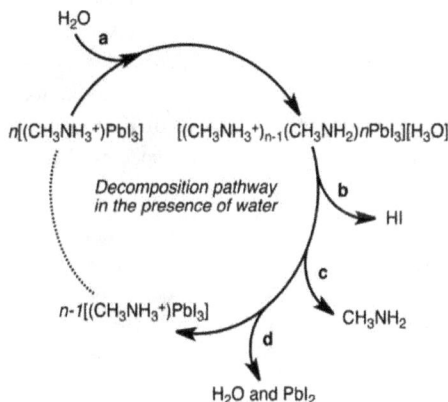

$$CH_3NH_3PbI_3 + nH_2O \rightarrow (CH_3NH_3)_4 PbI_6.2nH_2O \tag{6}$$

Figure 5. Reaction mechanism of the degradation of OMHP-SCs in the presence of water (Reproduce with copyright permission) [50].

5.3   Effect of temperature

The stability of OMHPs is also affected by the temperature of the environment. Mainly, temperature distorts the crystal structure and phase transition of the OMHPs-SCs. The crystal structure of MAPbBr$_3$ and MAPbCl$_3$ at room temperature presents a cubic structure but when the temperature is increased to 327.4 K, the shape of MAPbI3 is distorted into a tetragonal structure. A slight distortion of the PbI$_6$ octahedral around the c-axes also occurs. When the temperature is lowered it would result in a change in the crystallography of the crystal, in this case, the structure is changed into an orthorhombic phase from a tetragonal phase [52].

5.4   Effect of oxygen and light

From all types of OMHP-SCs, CH$_3$NH$_3$PbI$_3$ is one of the earliest PSCs having low-cost and solution-processed optoelectronics characteristics. It is revealed that the photoactive layers of CH$_3$NH$_3$PbI$_3$ exposed to light and oxygen lead to superoxide (O^{2-}) species formation. Superoxide is a highly reactive (O^{2-}) species and photo-excited MAPbI3 deprotonated by this superoxide and thus MA-cation (CH$_3$NH$_3^+$) and PbI$_2$ are formed. Transient absorption spectroscopy (TAS) which studied the interfacial CT of OMHPs-SCs, revealed that due to oxygen degradation, photo-induced charge carriers are decreased. The degradation of PSCs in the presence of oxygen and light is shown in Fig. 6 [53]. If we place the OMHPs-SCs to the environmental light and oxygen only for 48 hours, it results in a blue shift that ranges from 780−520 nm and the color of the sample fairly changes into yellow from the dark brown. Both of these findings make the PSCs less stable and lead to the breakdown of the film of perovskite. This data is taken from previous work but now with help of the passivation of PSCs, the film of perovskite is quite stable [54]. Oxidation of halogen ions also takes place due to ultraviolet illumination. So, ultraviolet illumination breaks Pb-X bonds and generates the free radicals of X-ions. Light energy also starts the irreversible reaction of MAPbI3 in which the stable species PbI$_2$ is formed [55].

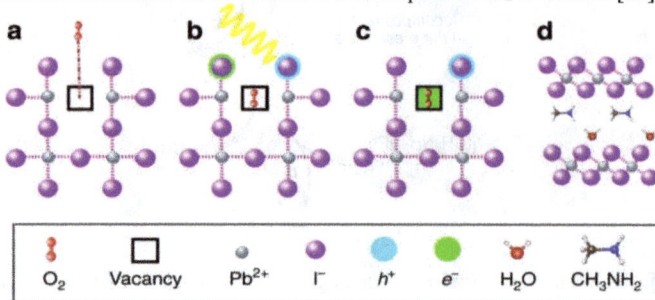

Figure 6. (a) Oxygen diffusion in the lattice of PSC. (b) Photo-excitation of MA and create electron and hole. (c) Formation superoxide. (d) Reaction and degradation of crystal (Reproduce with copyright permission) [53].

6.　　Recent development in the OMHP solar cells

6.1　　Ion migration and the suppression of ions

No doubt, the migration of charge ion carriers is one of the biggest problems in the stability of PSCs. The constituents (C-Si) of first-generation-based SCs are strongly attached to each other through covalent bond linkage. In contrast, the constituents of 3rd generation-based SCs (OMHPs) are weakly attached to each other through intermolecular and intramolecular forces like ionic bonds, H-bonding, and other Van der Waals forces. These forces (Ionic and hydrogen bonding etc.) easily combine with moisture, water, and oxygen. The components of OMHP-SCs e.g. $MAPbI_3$ (I^-, MA^+, and Pb^{2+}) are all larger in size therefore, its polarizability power is high and they become easily polarized to make perovskite a soft lattice as shown in Eqs. 7, 8 and 9 [56]. These characteristics make perovskite lattice less stable, eventually leading to the loss of photoelectric properties and ions migrating towards each other to form more stable molecules [57].

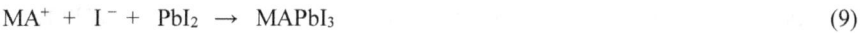

$$MAPbI_3 \rightarrow MA^+ + I^- + PbI_2 \tag{7}$$

$$2I^- \rightarrow I_{2(g)} + 2e^- \tag{8}$$

$$MA^+ + I^- + PbI_2 \rightarrow MAPbI_3 \tag{9}$$

Ion migration takes place in lead halide perovskites when either the A, B, or X element within the ABX_3 perovskite structure becomes a mobile ionic species and starts migrating through the perovskite layer. So the migration of ions is suppressed by using different techniques.

To overcome this problem, researchers are using cross-linking materials like polydimethylsiloxane (PDMS) to dope the perovskite-crystal layer in order to improve environmental stability. The doping of PDMS on OMHPs-SCs promoted the crystallization phase of perovskite. This doping also increases J_{SC} and FF values of SCs. This experiment indicates that PDMS doping with an appropriate amount of curing agent yields good environmentally stable OMHPs-SCs [58]. The suppression of the movement of ions and reduction in the SCs-crystal defects can be solved by perfluorotetradecanoic acid (PFTeDA) with a carbonyl unit and carbon-fluorine bonds. [59]. A quasi-two-dimensional (2.5D) perovskite is also helpful in the suppression of ion migration. Less than 1% relative PCE in over 80 h is lost when 2.5D PSCs are used. On the other hand, 3D PSCs efficiency drops by more than 50% in just the first 24 h [60].

Perovskite based Materials for Energy Storage Devices   Materials Research Forum LLC
Materials Research Foundations 151 (2023) 33-66   https://doi.org/10.21741/9781644902738-2

6.2   Solvent engineering

For the development of a uniform PSC-crystal structure, solvent-engineering technology is used. With the help of solvent engineering, we were able to develop extremely smooth perovskite layers. The main purpose of solvent engineering was to improve the PCE of OMHP-SCs by producing defects in the crystal of perovskite. Actually, a high density grain boundary defect is created in the crystal which renders the migration of ions and hence, the stability of PSCs is increased to a great extent. Therefore, solvent engineering technology uses the combination of solvent–anti solvent approach. The study of the synergistic effect of solvent and antisolvent is very helpful in the solubility of the precursors. Anti-solvent helps to obtain super-saturation. Super-saturation is the requisite for nucleation as well as the formation of the perovskite phase. The rate of formation of crystals affects the crystal grain size [61]. The intercalation process is used in the preparation of a stable MAI (Br)–PbI_2–DMSO perovskite. During this process, a non-dissolving solvent creates retardation in the fast reaction of $PbI(Br)_2$ and MAI(Br), this retardation enables the shaping of a particularly smooth and thick surface. For instance, when the combination of two solvents is used (γ-butyrolactone and DMSO) followed by toluene drop-casting, it leads to highly smooth and thick perovskite layers. By using this technique PCE of SCs increased up to 16.2%. Understanding the role of solution processing or solvent engineering technology provides a different way to develop highly well-organized SCs. It is a suitable strategy for the development of a stable PbI_2-based perovskite crystal layer through the intercalation process. This technique is helpful in the development of hetero-junction in OMHP-based SCs in the future [62].

6.3   Annealing

The process of annealing is another method to improve the stability and PCE of OMHPs-SCs. With a short post-annealing of less than 3 min (2.5 minutes) in an ambient environment, the film of perovskite is developed with a more concise stoichiometric calculation. The short time period of annealing produced more intrinsic perovskite films and saved energy as well. Annealing also encourages the spontaneous de-doping of perovskites during aging resulting in less charge recombination in layers and giving high PCE [63]. Out of two methods introduced for annealing, the first one alters the surface structure and electro-optic properties of perovskite film, and the second one develops highly smooth and well-crystalline perovskite SCs.

6.4   2D/3D technology

Two-dimensional (2D) and three-dimensional (3D) OMHPs-SCs exhibit excellent photoelectric properties. The crystal structures of 2D-3D OMHPs are shown in Fig. 7. The

2D-3D technology is regarded as an emerging solution for the stability of OMHP-SCs. The SCs based on this technology show long-period stability, restricted ion movement, large band gap, and efficient PCE. We have already discussed the degradation of PSCs with environmental components like light, oxygen, water, moisture, etc. The ability of 2D PSCs to be attached with bigger, less volatile cations which are hydrophobic in nature makes it superior to 3D PSCs. Therefore, it was seen that 2D base OMHPs-SCs are more stable as opposed to 3D-based SCs. The amalgam of 2D/3D-based SCs exhibits excellent stability in the environment. 2D PSCs with bifunctional ammonium cations, like $Y(CH_2)_2NH_3^+$ (Y, F, Cl, Br, I, CN), opened a new class of PV devices that have the ability to tunable bandgap. The CT mechanism is improved by using 2D PSCs, especially when the aromatic cation is used at the A-site of the ABX_3-crystal. Research is continuing to raise the stability and power of 2D PSCs. The unique properties of 2D-perovskite like ion migration, effective masses of holes and electrons, and conversion efficiency replace the 3D PSC [64].

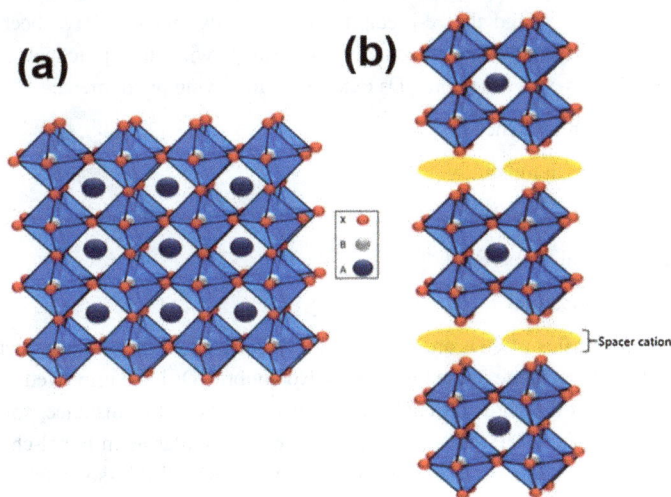

Figure 7. (a) It is a 3D based developed perovskite. (b) It is a 2D based developed perovskites (Reproduce with copyright permission) [64].

6.5   Organometallic halides-based perovskite quantum dot solar cells

Nanotechnology made progress in a lot of fields and its popularity and practical applications are increasing day by day. Nanotechnology-based OMHPs-SCs are opening a new concept in the area of SE. With the help of nanomaterials, the PCE and stability of SCs have increased. The problem of environmental stability and efficiency has also been solved to great extent. Nanostructured devices are considered inexpensive energy-absorbing devices. Quantum dots (QDs) are actually man-made nano-based crystals [65]. QDs are important in the field of PV devices due to the following reasons:

- They control the energy flow at an atomic level.
- Creation from inexpensive materials in an energy-saving environment (room temperature).
- No need to be purified repeatedly as in silicon-based (1st generation).
- Smooth CT mechanism.
- Impart a variety of colors, when the UV light hits them.

Different procedures for the preparation of QDs have been proposed. Usually, the classical hot-injection technique and the re-precipitation synthesis methods have been used for organic nano-crystals at ambient conditions. Along with the practical application perspective, OMHPs-SCs based on QDs exhibit the following phenomena:-

- High photoluminescence
- Wide wavelength tunability
- Ultra-narrow band emissions
- Superior optical properties
- Low-cost fabrication

Some unique properties of QDs are shown in Fig. 8 [66]. From the last two years, a lot of development has been made in the QDs-PSCs. No doubt, QDs have improved the stability and PCE of SCs but they are still challenging for researchers. For instance, solvent, light and heat stability of OMHP-SCs based on QDs are not similar as in metal-chalcogenide QDs therefore, it is mandatory to improve the quality of QDs OMHP-solar cells. However, the traditional QDs based devices are better than the devices based on OMHP-QDs [67].

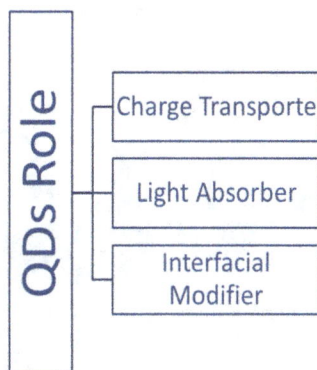

Figure 8. *The fundamental properties of quantum dots based perovskite are shown in this diagram (Self-drawn).*

6.6   Solid-state hole conductor-free (HCF) OMHP-SCs

The instability of OMHP-SCs is less in liquid electrolytes. To solve this problem, we use solid-state PSCs. In HCF-type SCs, spiro-OMeTAD is frequently used as a hole transport material (HTM). Comparatively stable PSCs are developed with the solid-state hole conductor PSCs [56]. Furthermore, scientists were able to increase the efficiency of HTM-free perovskite by using a 300 *nm* mesoporous TiO_2 film [68]. To solve the problem of air degradation of PSCs, different types of protective layers have been used. For example, a hydrophobic layer of buffer, the use of a water-resistant and mesoporous layer of TiO_2 and Al_2O_3, etc. These measures improve the stability of $MAPbI_3$. NiO2 is stable showing a high charge carrier mobility with a wide band gap (~3.6 eV). Nickel oxide has the ability to pull out the holes and intercept the electrons in PSCs. From this oxide, a new concept HCF-PSCs (hole-free SCs) was introduced [69]. Lioz Etgar and his group members reported the first HCF-OMHPs ($MAPbI_3/TiO_3$ hetero-junction) SCs as shown in Fig. 9. Their experiment showed the impressive result for this hetero-junction cell, $Jsc = 16.1$ mA/cm^2, $Voc = 0.631$ V and $FF = 0.57$ [70]. HCF perovskite extended the stability of PSCs but still improvements are needed.

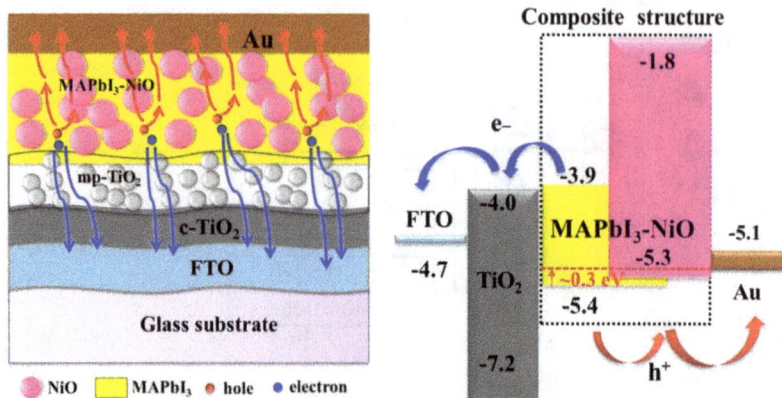

Figure 9. *Device fabrication of hole conductor free perovskite SCs with their energy levels diagram of MAPbI₃ with TiO₂ protective layer (Reproduced with copyright permission) [70].*

6.7 Tandem perovskite solar cells (TPSCs)

TPSCs were introduced to overcome the band gap imparting and the thermalization of hot charge carriers. The stack of *p-n* junctions in which each part of the junction is formed by semiconductors of different band gaps is known as a tandem cell. It shows a higher value of efficiency as compared to single-junction SCs [71]. Both types of Tandem-SCs have been developed. Organic tandem-SCs have some advantages over inorganic tandem-SCs. TPSCs performance rate was higher than the investigated perovskite and reached up to 25.2% efficiency [72].

6.8 Passivation of OMHP-SCs

In PV technology, passivation generally refers to any physical or chemical change introduced to the cell for its better performance. Today's research introduces a number of passivation methods or techniques for the stability and PCE of OMHPs-SCs. With the help of passivation techniques, intrinsic and extrinsic defects can be minimized. Passivation not only improves the V_{OC}, but also helps to enhance the environmental stability of OMHP-SCs. We have discussed a lot of methods above which are used to improve the quality of PSCs and help to make them commercialized. The passivation technique may occur with the perovskite bulk or at interfaces. Recent research reports revealed that by using passivation, the size of the crystal and grain boundaries can also be controlled [73].

Conclusion

The 21st century is facing highly considerable and critical issues like overpopulation, food shortage, environmental problems, global warming, and energy as well. From all of these, we have particularly discussed the worldwide energy crisis, with one of its best solutions, known as Solar Energy. To overcome energy challenges, different techniques and methods have been developed during the last 20 years. SE is one of the unique solutions which is sustainable, environment-friendly, low-cost, and simple. A lot of materials have been suggested by scientists for solar cell fabrications. Organometallic halides-based perovskite materials are promising. The solar cells made of these materials act as semiconductors and can convert light energy into electrical signals. Their efficiency rate is highly improved with different passivation techniques. They are comparable with silicon-based SCs with respect to efficiency rates but their cost is too low as compared to silicon-based SCs. Due to their degradability by oxygen, light, and temperature, they have not been commercialized on large scale yet. If we are able to overcome their environmental stability issues, they will rise as emerging solar cells in the world with low costs and high PCE.

References

[1] H.J. Snaith, Perovskites: The emergence of a new era for low-cost, high-efficiency solar cells, J. Phys. Chem. Lett. 4 (2013) 3623-3630. https://doi.org/10.1021/jz4020162

[2] L.M. Fraas, Low-cost Solar Electric Power, Springer, 2014. https://doi.org/10.1007/978-3-319-07530-3

[3] S.M. Hasnain, Solar energy education: A viable pathway for sustainable development, Renew. Energy 14 (1998) 387-392. https://doi.org/10.1016/S0960-1481(98)00094-9

[4] S. Mekhilef, S.Z. Faramarzi, R. Saidur, Z. Salam, The application of solar technologies for sustainable development of the agricultural sector, Renew. Sustain. Energy Rev. 18 (2013) 583-594. https://doi.org/10.1016/j.rser.2012.10.049

[5] I. Khan, F. Hou, H.P. Le, The impact of natural resources, energy consumption, and population growth on environmental quality: Fresh evidence from the United States of America, Sci. Total Environ. 754 (2021) 142222. https://doi.org/10.1016/j.scitotenv.2020.142222

[6] M.A. Aktar, M.M. Alam, A.Q.A. Amin, Global economic crisis, energy use, CO2 emissions, and policy roadmap amid COVID-19, Sustain. Prod. Consum. 26 (2021) 770-781. https://doi.org/10.1016/j.spc.2020.12.029

[7] A. Mikhaylov, Sustainable development and renewable energy: A new view to a global problem, Energies 15 (2022) 1397. https://doi.org/10.3390/en15041397

[8] M.V. Dambhare, B. Butey, S.V. Moharil, Solar photovoltaic technology: A review of different types of solar cells and its future trends, J. Phys. Conf. Ser. 1913 (2021) 012053. https://doi.org/10.1088/1742-6596/1913/1/012053

[9] D. Verma, S. Nema, A.M. Shandilya, S.K. Dash, Maximum Power Point Tracking (MPPT) techniques: Recapitulation in solar photovoltaic systems, Renew. Sustain. Energy Rev. 54 (2016) 1018-1034. https://doi.org/10.1016/j.rser.2015.10.068

[10] M. Kratschmann, E. Dütschke, Selling the sun: A critical review of the sustainability of solar energy marketing and advertising in Germany, Energy Res. Soc. Sci. 73 (2021) 101919. https://doi.org/10.1016/j.erss.2021.101919

[11] S.O. Abril, J.A.P. León, J.O.G. Mendoza, Study of the benefit of solar energy through the management of photovoltaic systems in Colombia, Int. J. Energy Econ. Policy. 11 (2021) 96-103. https://doi.org/10.32479/ijeep.10706

[12] T. Tsoutsos, N. Frantzeskaki, V. Gekas, Environmental impacts from solar energy technologies, Energy Policy. 33 (2005) 289-296. https://doi.org/10.1016/S0301-4215(03)00241-6

[13] M.H. Shakil, Z.H. Munim, M. Tasnia, S. Sarowar, COVID-19 and the environment: A critical review and research agenda, Sci. Total Environ. 745 (2020) 141022. https://doi.org/10.1016/j.scitotenv.2020.141022

[14] International Energy Outlook 2021. https://www.eia.gov/outlooks/ieo/

[15] B.P. Singh, S.K. Goyal, P. Kumar, Solar PV cell materials and technologies: Analyzing the recent developments, Mater. Today Proc. 43 (2021) 2843-2849. https://doi.org/10.1016/j.matpr.2021.01.003

[16] J. Bisquert, The Physics of Solar Energy Conversion, first ed., CRC Press, 2022.

[17] M.A. Green, Photovoltaic principles, Physica E: Low-Dimens. Syst. Nanostruct. 14 (2002) 11-17. https://doi.org/10.1016/S1386-9477(02)00354-5

[18] W.A. Badawy, A review on solar cells from Si-single crystals to porous materials and quantum dots, J. Adv. Res. 6 (2015) 123-132. https://doi.org/10.1016/j.jare.2013.10.001

[19] M.A. Green, Third generation photovoltaics: Solar cells for 2020 and beyond, Physica E: Low-Dimens. Syst. Nanostruct. 14 (2002) 65-70. https://doi.org/10.1016/S1386-9477(02)00361-2

[20] G.H. Kim, D.S. Kim, Development of perovskite solar cells with >25% conversion efficiency, Joule 5 (2021) 1033-1035. https://doi.org/10.1016/j.joule.2021.04.008

[21] P. Roy, N.K. Sinha, S. Tiwari, A. Khare, A review on perovskite solar cells: Evolution of architecture, fabrication techniques, commercialization issues and status, Sol. Energy. 198 (2020) 665-688. https://doi.org/10.1016/j.solener.2020.01.080

[22] C.D. Bailie, M.D. McGehee, High-efficiency tandem perovskite solar cells, MRS Bull. 40 (2015) 681-685. https://doi.org/10.1557/mrs.2015.167

[23] M. Mohan, Life cycle assessment, in: A. Thankappan, S. Thomas (Eds.), Perovskite photovoltaics: Basic to Advanced Concepts Implementation, Elsevier Inc., 2018, pp. 447-480. https://doi.org/10.1016/B978-0-12-812915-9.00014-9

[24] S.D. Wolf, J. Holovsky, S.J. Moon, P. Löper, B. Niesen, M. Ledinsky, F.J. Haug, J.H. Yum, C. Ballif, Organometallic halide perovskites: Sharp optical absorption edge and its relation to photovoltaic performance, J. Phys. Chem. Lett. 5 (2014) 1035-1039. https://doi.org/10.1021/jz500279b

[25] M.A. Green, A.H. Baillie, H.J. Snaith, The emergence of perovskite solar cells, Nat. Photonics 8 (2014) 506-514. https://doi.org/10.1038/nphoton.2014.134

[26] X. Wang, M. Li, B. Zhang, H. Wang, Y. Zhao, B. Wang, Recent progress in organometal halide perovskite photodetectors, Org. Electron. 52 (2018) 172-183. https://doi.org/10.1016/j.orgel.2017.10.027

[27] N. Kumar, J. Rani, R. Kurchania, A review on power conversion efficiency of Lead Iodide perovskite-based solar cells, Mater. Today Proc. 46 (2020) 5570-5574. https://doi.org/10.1016/j.matpr.2020.09.349

[28] N. Ashurov, B.L. Oksengendler, S. Maksimov, S. Rashiodva, A.R. Ishteev, D.S. Saranin, I.N. Burmistrov, D. V. Kuznetsov, A.A. Zakhisov, Current state and perspectives for organo-halide perovskite solar cells. Part 1. Crystal structures and thin film formation, morphology, processing, degradation, stability improvement by carbon nanotubes: A review, Mod. Electron. Mater. 3 (2017) 1-25. https://doi.org/10.1016/j.moem.2017.05.001

[29] M. Azzouzi, T. Kirchartz, J. Nelson, Factors controlling open-circuit voltage losses in organic solar cells, Trends Chem. 1 (2019) 49-62. https://doi.org/10.1016/j.trechm.2019.01.010

[30] D. Kiermasch, L.G. Escrig, H.J. Bolink, K. Tvingstedt, Effects of masking on open-circuit voltage and fill factor in solar cells, Joule. 3 (2019) 16-26. https://doi.org/10.1016/j.joule.2018.10.016

[31] E. Mosconi, P. Umari., F.D. Angelis, Electrical and optical properties of MAPBX3 perovskites (X = I, Br, Cl): A unified DFT and GW theoretical analysis, Phys. Chem. Chem. Phys. 18 (2016) 27158-27164. https://doi.org/10.1039/C6CP03969C

[32] Y. Dang, D. Ju, L. Wang, X. Tao, Recent progress in the synthesis of hybrid halide perovskite single crystals, CrystEngComm. 18 (2016) 4476-4484. https://doi.org/10.1039/C6CE00655H

[33] C.J. Bartel, C. Sutton, B.R. Goldsmith, R. Ouyang, C.B. Musgrave, L.M. Ghiringhelli, M. Scheffler, New tolerance factor to predict the stability of perovskite oxides and halides, Sci. Adv. 5 (2019) 1-10. https://doi.org/10.1126/sciadv.aav0693

[34] F. Zhang, K. Zhu, Additive engineering for efficient and stable perovskite solar cells, Adv. Energy Mater. 10 (2019) 1902579. https://doi.org/10.1002/aenm.201902579

[35] D. Ji, S.Z. Feng, L. Wang, S. Wang, M. Na, H. Zhang, C.M. Zhang, X. Li, Regulatory tolerance and octahedral factors by using vacancy in APbI3 perovskites, Vacuum. 164 (2019) 186-193. https://doi.org/10.1016/j.vacuum.2019.03.018

[36] Y.C. Hsiao, T. Wu, M. Li, Q. Liu, W. Qin, B. Hu, Fundamental physics behind high-efficiency organo-metal halide perovskite solar cells, J. Mater. Chem: A, 3 (2015) 15372-15385. https://doi.org/10.1039/C5TA01376C

[37] G.R. Berdiyorov, F.E. Mellouhi, M.E. Madjet, F.H. Alharbi, F.M. Peeters, S. Kais, Effect of halide-mixing on the electronic transport properties of organometallic perovskites, Sol. Energy Mater. Sol. Cells. 148 (2016) 2-10. https://doi.org/10.1016/j.solmat.2015.11.023

[38] X. Liu, W. Zhao, H. Cui, Y. Xie, Y. Wang, T. Xu, F. Huang, Organic-inorganic halide perovskite based solar cells - revolutionary progress in photovoltaics, Inorg. Chem. Front. 2 (2015) 315-335. https://doi.org/10.1039/C4QI00163J

[39] L. Yang, A.T. Barrows, D.G. Lidzey, T. Wang, Recent progress and challenges of organometal halide perovskite solar cells, Reports Prog. Phys. 79 (2016) 26501. https://doi.org/10.1088/0034-4885/79/2/026501

[40] S. Bai, Z. Wu, X. Wu, Y. Jin, N. Zhao, Z. Chen, Q. Mei, X. Wang, Z. Ye, T. Song, R. Liu, S.T. Lee, B. Sun, High-performance planar heterojunction perovskite solar cells: Preserving long charge carrier diffusion lengths and interfacial engineering, Nano Res. 7 (2014) 1749-1758. https://doi.org/10.1007/s12274-014-0534-8

[41] L. Etgar, Hole Conductor Free Perovskite-based Solar Cells, Springer, 2016. https://doi.org/10.1007/978-3-319-32991-8

[42] G. Giorgi, J. Fujisawa, H. Segawa, K. Yamashita, Small photocarrier effective masses featuring ambipolar transport in Methylammonium Lead Iodide perovskite: A Density Functional Analysis, J. Phys. Chem. Lett. 4 (2013) 4213-4216. https://doi.org/10.1021/jz4023865

[43] U. Mehmood, A.A. Ahmed, M. Afzaal, F.A.A. Sulaiman, M. Daud, Recent progress and remaining challenges in organometallic halides based perovskite solar cells, Renew. Sustain. Energy Rev. 78 (2017) 1-14. https://doi.org/10.1016/j.rser.2017.04.105

[44] F. Paquin, J. Rivnay, A. Salleo, N. Stingelin, C. Silva, Multi-phase semicrystalline microstructures drive exciton dissociation in neat plastic semiconductors, J. Mater. Chem: C, 3 (2015) 10715-10722. https://doi.org/10.1039/C5TC02043C

[45] I. Borriello, G. Cantele, D. Ninno, Ab initio investigation of hybrid organic-inorganic perovskites based on Tin halides, Phys. Rev. B - Condens. Matter Mater. Phys. 77 (2008) 1-9. https://doi.org/10.1103/PhysRevB.77.235214

[46] F. Brivio, A.B. Walker, A. Walsh, Structural and electronic properties of hybrid perovskites for high-efficiency thin-film photovoltaics from first-principles, APL Mater. 1 (2013) 14-19. https://doi.org/10.1063/1.4824147

[47] H. Zarenezhad, T. Balkan, N. Solati, M. Halali, M. Askari, S. Kaya, Efficient carrier utilization induced by conductive polypyrrole additives in organic-inorganic halide perovskite solar cells, Sol. Energy. 207 (2020) 1300-1307. https://doi.org/10.1016/j.solener.2020.07.059

1 [48] F. Zhang, K. Zhu, On-device lead sequestration for perovskite solar cells, Nature 578 (2020) 555-558. https://doi.org/10.1038/s41586-020-2001-x

[49] M.M. Byranvand, A.N. Kharat, N. Taghavinia, Moisture stability in nanostructured perovskite solar cells, Mater. Lett. 237 (2019) 356-360. https://doi.org/10.1016/j.matlet.2018.10.029

[50] D. Wang, M. Wright, N.K. Elumalai, A. Uddin, Stability of perovskite solar cells, Sol. Energy Mater. Sol. Cells. 147 (2016) 255-275. https://doi.org/10.1016/j.solmat.2015.12.025

[51] J.A. Christians, P.A.M. Herrera, P.V. Kamat, Transformation of the excited state and photovoltaic efficiency of CH3NH3PbI3 perovskite upon controlled exposure to humidified air, J. Am. Chem. Soc. 137 (2015) 1530-1538. https://doi.org/10.1021/ja511132a

[52] A. Poglitsch, D. Weber, Dynamic disorder in Methylammonium trihalogenoplumbates (II) observed by millimeter-wave spectroscopy, J. Chem. Phys. 87 (1987) 6373-6378. https://doi.org/10.1063/1.453467

[53] N. Aristidou, C. Eames, I.S. Molina, X. Bu, J. Kosco, M.S. Islam, S.A. Haque, Fast oxygen diffusion and Iodide defects mediate Oxygen-induced degradation of perovskite solar cells, Nat. Commun. 8 (2017) 1-10. https://doi.org/10.1038/ncomms15218

[54] D. Bryant, N. Aristidou, S. Pont, I.S. Molina, T. Chotchunangatchaval, S. Wheeler, J.R. Durrant, S.A. Haque, Light and Oxygen induced degradation limits the operational stability of Methylammonium Lead triiodide perovskite solar cells, Energy Environ. Sci. 9 (2016) 1655-1660. https://doi.org/10.1039/C6EE00409A

[55] W. Chi, S.K. Banerjee, Development of perovskite solar cells by incorporating quantum dots, Chem. Eng. J. 426 (2021) 131588. https://doi.org/10.1016/j.cej.2021.131588

[56] M. Jeevaraj, S. Sudhahar, M.K. Kumar, Evolution of stability enhancement in organo-metallic halide perovskite photovoltaics-A review, Mater. Today Commun. 27 (2021) 102159. https://doi.org/10.1016/j.mtcomm.2021.102159

[57] M.G. Rosell, A. Bou, J.A.J. Tejada, J. Bisquert, P.L. Varo, Analysis of the influence of selective contact heterojunctions on the performance of perovskite solar cells, J. Phys. Chem: C, 122 (2018) 13920-13925. https://doi.org/10.1021/acs.jpcc.8b01070

[58] J. Kang, R. Huang, S. Guo, G. Han, X. Sun, I. Ismail, C. Ding, F. Li, Q. Luo, Y. Li, C.Q. Ma, Suppression of ion migration through cross-linked PDMS doping to enhance the operational stability of perovskite solar cells, Sol. Energy. 217 (2021) 105-112. https://doi.org/10.1016/j.solener.2021.01.025

[59] T. Ye, Y. Hou, A. Nozariasbmarz, D. Yang, J. Yoon, L. Zheng, K. Wang, K. Wang, S. Ramakrishna, S. Priya, Cost-effective high-performance charge-carrier-transport-layer-free perovskite solar cells achieved by suppressing ion migration, ACS Energy Lett. 6 (2021) 3044-3052. https://doi.org/10.1021/acsenergylett.1c01186

[60] Z. Huang, A.H. Proppe, H. Tan, M.I. Saidaminov, F. Tan, A. Mei, C.S. Tan, M. Wei, Y. Hou, H. Han, S.O. Kelley, E.H. Sargent, Suppressed ion migration in reduced-dimensional perovskites improves operating stability, ACS Energy Lett. 4 (2019) 1521-1527. https://doi.org/10.1021/acsenergylett.9b00892

[61] Z. Arain, C. Liu, Y. Yang, M. Mateen, Y. Ren, Y. Ding, X. Liu, Z. Ali, M. Kumar, S. Dai, Elucidating the dynamics of solvent engineering for perovskite solar cells, Sci. China Mater. 62 (2019) 161-172. https://doi.org/10.1007/s40843-018-9336-1

[62] N.J. Jeon, J.H. Noh, Y.C. Kim, W.S. Yang, S. Ryu, S. Seok, Solvent engineering for high-performance inorganic-organic hybrid perovskite solar cells, Nat. Mater. 13 (2014) 897-903. https://doi.org/10.1038/nmat4014

[63] Y. Deng, Z. Ni, A.F. Palmstrom, J. Zhao, S. Xu, C.H.V. Brackle, X. Xiao, K. Zhu, J. Huang, Reduced self-doping of perovskites induced by short annealing for efficient solar modules, Joule 4 (2020) 1949-1960. https://doi.org/10.1016/j.joule.2020.07.003

[64] E.B. Kim, M.S. Akhtar, H.S. Shin, S. Ameen, M.K. Nazeeruddin, A review on two-dimensional (2D) and 2D-3D multidimensional perovskite solar cells: Perovskites structures, stability, and photovoltaic performances, J. Photochem. Photobiol. C Photochem. Rev. 48 (2021) 100405. https://doi.org/10.1016/j.jphotochemrev.2021.100405

[65] M.C. Mathpal, P. Kumar, F.H. Aragón, M.A.G. Soler, H.C. Swart, Basic concepts, engineering, and advances in dye-sensitized solar cells, in S.K. Sharma, K. Ali (Eds.), Solar cells: From Materials to Device Technology, Springer, 2020, pp. 185-233. https://doi.org/10.1007/978-3-030-36354-3_8

[66] Y. Yang, W. Wang, Effects of incorporating PbS quantum dots in perovskite solar cells based on CH3NH3PbI3, J. Power Sources. 293 (2015) 577-584. https://doi.org/10.1016/j.jpowsour.2015.05.081

[67] G.L. Yang, H.Z. Zhong, Organometal halide perovskite quantum dots: Synthesis, optical properties, and display applications, Chinese Chem. Lett. 27 (2016) 1124-1130. https://doi.org/10.1016/j.cclet.2016.06.047

[68] N.G. Park, Perovskite solar cells: An emerging photovoltaic technology, Mater. Today. 18 (2015) 65-72. https://doi.org/10.1016/j.mattod.2014.07.007

[69] F. Paquin, J. Rivnay, A. Salleo, N. Stingelin, C. Silva, Multi-phase semicrystalline microstructures drive exciton dissociation in neat plastic semiconductors, J. Mater. Chem: C, 3 (2015) 10715-10722. https://doi.org/10.1039/C5TC02043C

[70] Y. Wang, W. Rho, H. Yang, T. Mahmoudi, S. Seo, D. Lee, Y. Hahn, Air-stable, hole-conductor-free high photocurrent perovskite solar cells with CH3NH3PbI3-NiO nanoparticles composite, Nano Energy. 27 (2016) 535-544. https://doi.org/10.1016/j.nanoen.2016.08.006

[71] N.N. Lal, Y. Dkhissi, W. Li, Q. Hou, Y.B. Cheng, U. Bach, Perovskite tandem solar cells, Adv. Energy Mater. 7 (2017) 1-18. https://doi.org/10.1002/aenm.201602761

[72] T. Todorov, T. Gershon, O. Gunawan, C. Sturdevant, S. Guha, Perovskite-kesterite monolithic tandem solar cells with high open-circuit voltage, Appl. Phys. Lett. 105 (2014) 173902. https://doi.org/10.1063/1.4899275

[73] F. Wang, S. Bai, W. Tress, A. Hagfeldt, F. Gao, Defects engineering for high-performance perovskite solar cells, NPJ Flex. Electron. 2 (2018) 1-14. https://doi.org/10.1038/s41528-017-0014-9

Materials Research Forum LLC
https://doi.org/10.21741/9781644902738-3

Chapter 3

Perovskite Based Ferroelectric Materials for Energy Storage Devices

M. Rizwan[1*], A. Ayub[2], T. Fatama[1], H. Hameed[1], Q. Ali[1], K. Aslam[1], T. Hashmi[1]

[1]School of Physical Sciences, University of the Punjab, Lahore, Pakistan

[2]Department of Physics, University of the Punjab, Lahore, Pakistan

*rizwan.sps@pu.edu.pk

Abstract

World's energy crisis has led to scrupulous research in the field of energy harvesting. Ferroelectrics have become the elite choice for energy storage applications such as in capacitors, transducers and sensors owing to their exciting properties such as ferroelectricity, remnant polarization and dielectric properties and high conversion efficiencies. The objective of this chapter is to elaborate the energy storage properties of ferroelectric perovskites as it is a necessity to construct such devices to meet the increasing demand of energy renewable resources. Lead based and lead-free ferroelectrics and various ferroelectric based energy storage devices as well as the ways to optimize their energy storage density are meticulously discussed.

Keywords

Ferroelectricity, Capacitors, Dielectrics, High Storage Density, Fuel Cells, Perovskite Solar Cells (PSCs), Transport Properties

Contents

Perovskite Based Ferroelectric Materials for Energy Storage Devices.......67

1.   Introduction...68

2.   Ferroelectricity..70

3.   Ferroelectric Perovskites 71

4.   Lead-Based Perovskite Ferroelectrics 72

        4.1   Niobate-Based Ferroelectrics 72

4.2   Lanthanum Based Ferroelectrics 73

4.3   Lead-Free Perovskite Ferroelectrics...73

4.3.1 Barium Titanate Based Ferroelectric...74

4.3.2 Alkaline Niobate Based Ferroelectric..74

4.3.3 Bismuth Based Ferroelectrics 75

5.   **Energy Storage Devices**..**75**

5.1   Types of Energy Storage Devices 76

5.1.1 Battery Energy Storage..76

5.1.2 Thermal Energy Storage 76

5.1.3 Pumped Hydroelectric Energy Storage 76

5.1.4 Mechanical Energy Storage...77

5.1.5 Hydrogen Energy Storage 77

6.   **Transport Properties**..**78**

7.   **Energy Density of Ferroelectrics**..**78**

7.1   Ways to Improve Energy Density 79

7.1.1 Chemical Substitution..79

8.   **High Energy Efficiency Perovskite Solar Cells** **80**

9.   **Ferroelectrics for Energy Storage Devices**...**81**

9.1   Fuel Cells..82

9.2   Photocatalysts 83

9.2.1 Characterization and Preparation of Photo Catalysts.....................................83

9.3   Capacitive Energy Storage Devices 84

Conclusion..**84**

References **85**

1.   Introduction

The excessive use of electronic devices in the modern era is causing an increase in energy usage, which needs to be sought out by using advanced energy storage devices. In recent years, there has been a focus towards low-cost renewable energy sources and it is being commended globally. Silicon has been used as a primary semiconductor device since the

1950's. However, the use of silicon crystals utilizes a lot of energy and it is expensive too. In search for an alternative, the use of perovskites has become liable as they have the same semiconducting properties as silicon. The vigorous optical and magnetic properties and applications in the late 20th century introduced the broad family of perovskite materials [1,2].

Perovskite is actually a cubic crystalline semiconductor alike calcium titanate mineral with the general formula $CaTiO_3$. The discovery of the mineral took place in the Ural Mountains of Russia by Gustav Rose in 1839 and the origin of its name came from the Russian Mineralogist Lev Perovski (1792 – 1856). The general chemical formula usually dedicated to perovskites is ABX_3 where A & B are cations, X is an anion, usually oxide of halogen. Perovskites are one of the astonishing materials of the 21st century. Ever since their discovery, there has been immense research in the field of perovskites due to their prospective applications in energy storage, optoelectronics devices as well as in pollutant degradation, due to exceptional photoelectric and catalytic properties [3]. They use light energy to transport electric charge which is one of their fundamental and useful characteristics. Perovskites are found in the Earth's mantle. They are found as tiny anhedral to subhedral crystal fillings, spaces between the rock-forming silicates [4-6].

Perovskites are replacing conventional materials in energy storage devices and there are many perovskites being used in energy storage applications. Such as organic-inorganic perovskites are being researched for applications in solar cells. The main energy storage devices are batteries, solar cells, fuel cells, and capacitors. Dielectric capacitors are preferred due to their temperature stability. Dielectric perovskites are being researched for capacitors. Ferroelectric materials such as barium titanate and relaxer ferroelectric materials are useful for energy storage capacitors. The unique properties of perovskite materials have made them the most fundamental and important material for optoelectronic applications. Perovskites are formed by the combination of a large number of elements. Thus, more efficient structures can be made. The greater the combination of elements, the more will be the possibility of having an efficient material. This factor is also helpful in making these materials, low-cost energy materials for photonics devices. Since a large number of different elements are directly involved in perovskites, thus these materials are the most extensively studied materials [7].

The confined structure and quantum-sized nature of perovskites make them useful for field effect transistors (FET) and light-emitting diodes (LEDs) applications. These materials are widely used in photo detectors, nano-lasers, and waveguides. Perovskites are widely used as solar cells which yield a good performance and are liable. Perovskite crystals are also found in ultrasonic machines and memory chips. The unique properties of perovskites make them suitable for photovoltaic applications [3]. By using these materials, the efficiency of photovoltaic devices has increased from 3.8 to 22.1 %.

The grain size, uniformity, confined structure, and morphological qualities of perovskites help to determine the performance of devices. Perovskites are abundant and stable which means that these are found easily in nature and are not reactive to the environment [7].

Perovskite materials have gained much attention in semiconductor and optoelectronic devices. Since all perovskites are based on semiconductors, thus they are not an environmental hazard and are safe to use. Perovskites use light energy and convert this energy into electrical energy. Thus, these materials can be used for clean energy purposes. Moreover, the cost of materials is not so high, since natural materials are being used. Perovskites are abundant and stable which means that these are available easily in nature and are not reactive to the environment.

Energy storage solutions need to be effective for successful application as renewable energy sources. Dielectric ceramics are in high demand due to their properties such as high dielectric constant and storage density. The most reliable contenders for high-power energy storage applications are dielectrics that possess medium dielectric constant, low polarization hysteresis, and large breakdown strength. Some of these materials are lead-free perovskites. Hence, energy can be stored more efficiently by using perovskites [8].

2. Ferroelectricity

Ferroelectricity is a property exhibited by those materials that possess reversible electric spontaneous polarization on exposure to an external electric field [8]. The concept of ferroelectricity was given by Joseph Valasek in 1930 [9]. Ferroelectrics have small conductivity; therefore, they can be considered a special type of insulator. Ferroelectrics have eccentric pyroelectricity and piezoelectricity [10].

Ferroelectrics show non-zero polarization even if the external electric field is zero. In contrast to other dielectrics, the slope of the polarization curve of ferroelectric materials is nonlinear. The P-E characteristics of Ferroelectrics become non-linear and hysteretic [11]. The properties of ferroelectrics have rendered them significant materials for various applications. As the shape of the polarization curve of ferroelectrics is non-linear, thus ferroelectrics can be utilized in linear semiconductor devices such as resonators and high-k dielectrics. Ferroelectrics also have been used as microwave devices such as varactors or phase shifters. The non-linear characteristic of ferroelectric materials can be used as FRAM [11].

Perovskite based Materials for Energy Storage Devices   Materials Research Forum LLC
Materials Research Foundations 151 (2023) 67-88   https://doi.org/10.21741/9781644902738-3

Ferroelectric materials are excessively used in commercial and industrial applications, such as dielectric capacitors, electric sensors, and transducers. Perovskite layer structured ferroelectrics have better properties thus they serve greater progress in the studies of ferroelectrics [12]. The nonlinear character of ferroelectrics can be utilized in making capacitors. Ferroelectrics can be used as a memory functioning device due to their hysteresis function [11]. Ferroelectrics can also be used in sensor applications and ultrasonic machines. Ferroelectric tunnel junction (FTJ) is a slightly new idea in which a contact is made up of ferroelectric thick films that are situated between two electrodes [13]. Ferroelectrics can also be used to perform catalytic functions in various chemical processes [14].

3.   Ferroelectric Perovskites

Our future is endangered due to increased global warming. $CO2$ emissions have increased in the Earth's atmosphere due to the enormous amount of burning fuels. Not only this causes a rise in the temperature of our planet but also causes the reduction of natural energy resources. To compensate, we need to synthesize renewable energy storage devices which are of low cost and environment-friendly. Ferroelectric perovskites are among the contenders of suggested materials for such devices [15].

The energy storage devices consist of fuel cells, solar cells, and electrochemical capacitors along with dielectric capacitors [16]. Energy storage performance is determined by energy efficiency (η), and recoverable energy density (U_{erc}). These two factors are highly dependent on remnant polarization; breakdown strength and saturation polarization of material. Ferroelectric perovskites are widely used as energy storage devices, especially in ceramic capacitors. They have specific polarization, with a tailorable orientation by varying the direction of the applied exterior field [16].

They have zero polarization as a whole because equal numbers of domains are randomly oriented. As the externally applied field increases, cations attain enough energy so that the local potential barrier is swamped by domains, switching of domains is a result of the jumping of domains from positions of a random potential well to another allowed well position which is closely associated with the applied field. Polarization (Ps) is exhibited when E_{max} is applied; domain saturation state is a consequence of switching. It means no more reorientation is possible in the direction of the field. New domains can be created in opposite directions after reducing and retreating E which is a converse process.

P_r is the polarization which is non-zero in ferroelectrics, appearing at zero fields after reducing electric field. E_c, is the coercive field which is required to change the domains of ferroelectric materials back and forth. Once this process is polled, hysteresis loop followed by material will come back to zero, along with net polarization which will go back to -E_c, or on the other hand if the material is upturned above curie temperature T_c, but not at E=0 [8].

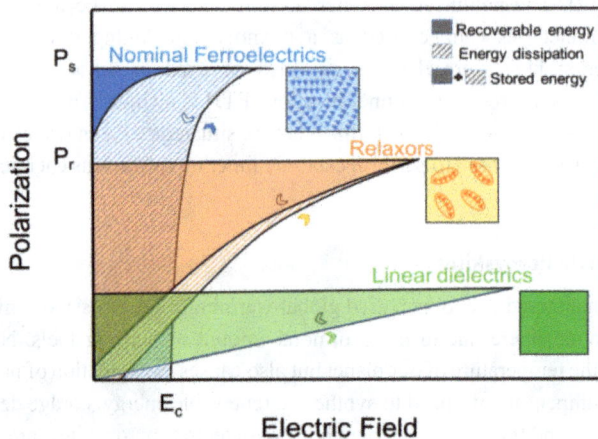

Figure 1. Polarization vs electric field (P-E) for typical ferroelectrics, relaxers, and linear dielectrics [8]

4. Lead-Based Perovskite Ferroelectrics

Over the past decades Pb based perovskite ferroelectrics have been investigated due to their progressive applications in capacitors, sensors, high-frequency transducers, large strain actuators, energy harvesters, and non-volatile memory devices. Outstanding ferroelectric and piezoelectric properties expressed by Pb-based lead zirconate solid solutions have vast applications in electronic devices and perovskite relaxers. Still, they have lower energy storage density than polymers because of low dielectric breakdown strength, which ultimately limits their energy storage applications [17]. Some of the perovskite-based ferroelectrics are discussed below.

4.1 Niobate-Based Ferroelectrics

Outstanding performance is exhibited by niobate-based ferroelectrics such as $(PbMg_{1/3}Nb_{2/3}O_3)(PbTiO_3)$ PMN-PT and $(PbZr_{1/3}Nb_{2/4}O_3)(PbTiO_3)$ PZN-PT close to MPB (morphotropic phase boundary) as they are the significant representations of relaxer based perovskites. PZT ($PbZr_{1-x}Ti_xO_3$) gained huge attention because of its excellent piezoelectric coefficients having a range of (d_{33}~200–750 pC N^{-1}) and high T_c (T_c~180–320 °C) and being used in several electronics. These compositions have secured a place in material science engineering due to their excellent energy storage applications. Because these solid solutions have a high strain response so they can be used in actuators, sensors, and transducers. Their low phase transition temperatures are responsible for their use in energy storage devices. Their dielectric properties are reduced above TRT ¼ 90–150 °C [18].

4.2 Lanthanum Based Ferroelectrics

(Pb, La)(Zr, Ti)O_3(PLZT) has gained enthusiasm as an A-site substitution material for Zr and Ti in PZT ceramic relaxor-ferroelectric material for energy storage devices. It is one of the members of the perovskite family which is well-known in many applications such as microwave devices, nonvolatile RAM, and photomechanical or electromechanical transducers. PLZT is being investigated meticulously because of its low P_r and high permittivity. (PLZT 9/65/35) film fabricated on a buffered layer made of platinum was reported by a group of scientists, with a value of energy density of 28.7 J/cc along with aerial capacitance density of 925 nF/cm^2 for 1 μm thick film. (PLZT 8/52/48) film placed on a buffer layer made of Lanthanum nitrate enclosed by nickel substrate (LNO/Ni) was proclaimed by Tong et al. with the value of highest energy density of 22J/cc and electrical energy storage efficiency of 77% % for a 3 μm film thickness. High energy capabilities are shown by these results approached to the electrochemical capacitors [19].

4.3 Lead-Free Perovskite Ferroelectrics

For decades, lead-based ferroelectrics have been utilized in piezoelectric applications, such as in Pb(Zr, Ti)O_3(PZT) and its offshoots. This is due to the promising piezoelectric properties, low cost, composition flexibility, and well-explained processing. However, lead is not safe for the environment and for humans as well, specifically in producing and recirculation of these materials. Due to recent awareness of environmental crises, alternatives of lead-based materials have been synthesized. Scientists have been putting all their efforts into producing ferroelectric materials having advanced electrochemical responses mainly used in transducers, actuators and sensors, and energy storage capacitors [20].

Perovskite structures that are lead-free FC (ferroelectric crystalized) are divided mainly into three groups: alkali bismuth titanates, alkaline niobates, and barium titanates. They show wide P–E loops with large P_r and E_c but small P_s–P_r, resulting in low U_{rec} and η. All three types possess enormous properties under certain specific conditions, but sometimes they could not show the flexibility of decent performance over a wide range of temperatures. $BaTiO_3$ at room temperature has a large piezoelectric coefficient but at 50 to 120^0C, it gets depolarized. Alkaline niobates, when compared to PZT show lower functional properties although they have higher Curie temperatures. However, many lead-free materials have been designed that are stated to accommodate industries [21].

4.3.1 Barium Titanate Based Ferroelectric

The perovskite structure of barium titanate BT has been broadly explored as lead-free ferroelectrics in dielectric capacitors. Bismuth titanate energy storage properties can be upgraded by enlarging the gap among P_r (remanent polarization), P_{max} (maximum polarization), and BDS (breakdown strength) by the formula [9]:

$$W_R = \int E \, dP \, P_{max} \qquad\qquad\qquad\qquad (Eq.1)$$

The other technique to tailor the energy stowing properties of BT is ion substitution i.e. A-site or B-site ion substitution; this makes BT a promising candidate for energy storage applications. It has recently been proclaimed that due to the same lone pair configuration $(6s^2)$ between lead and bismuth, BT can be used as an alternative to the lead-based ferroelectrics. BT ($BaTiO_3$) and BS ($BiScO_3$), when operated in a wide range of temperatures, can be favorable candidates for high energy density capacitors [21].

4.3.2 Alkaline Niobate Based Ferroelectric

$(K_{1-x}Na_x)NbO_3$ termed as sodium potassium niobate has gained interest since early 2000 for being lead-free PZ material. The pseudo-binary equilibrium phase of $KNbO_3$ and $NaNbO_3$ was first reported between 1960 to 1970. Comprehensive solid stability is exhibited by the binaries of these materials. At room temperature $NaNbO_3$ crystalizes in the orthorhombic syngony phase (Pbma). $NaNbO_3$ is well known among perovskites for having a higher number of phase transitions, although it is a prototype antiferroelectric. Orthorhombic (Amm_2) symmetry is exhibited by $KNbO_3$ at room temperature and is valid for all $(K_{1-x}Na_x)NbO_3$ (KNN) solids with a minimum 2% mole concentration of $KNbO_3$. Its composition which is thoroughly studied for PZ application is K/Na having a molar ratio of 1:1 and $K_{0.5} Na_{0.5} NbO_3$. At lower temperatures, it shows the phase of a rhombohedral (R_3c), between temperatures $16°C$ to $20°C$ it is orthorhombic and at $40°C$ it is tetragonal (P_{4mm}) and at higher temperatures, it is cubic (P_{m3m}). At room temperature, it has a monoclinic structure (P_m) [21].

4.3.3 Bismuth Based Ferroelectrics

BFO (bismuth ferrite) is a ferroelectric material (lead-free) that is attractive due to its electronic configuration (Bi^{3+}) that is quite similar to lead (Pb^{2+}). High curie temperature (~830⁰C) and large ferroelectric spontaneous polarization (~100µC/cm²) is exhibited by distorted perovskite structure (R3c) of BFO. The semiconducting nature of BFO is achieved at room temperature with a lower band gap (2.7eV) [21].

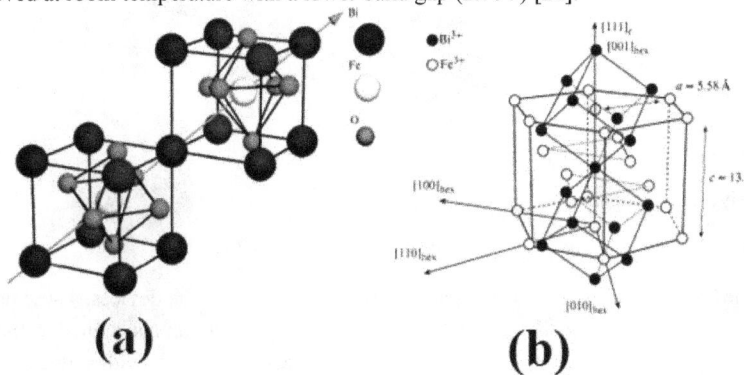

Figure 2. *Crystal Structure of BFO, a) Rhombohedral Structure b) Hexagonal Structure [21]*

BFO has attracted researchers because of its magnetic ordering and high-temperature ferroelectric nature. It is widely been used in ferroelectric tunneling junctions, magneto-electric RAM, non-volatile ferroelectric-based RAM, actuators as well as sensors [21].

5. Energy Storage Devices

Energy storage devices are systems that store energies like kinetic, potential, thermal, pressure, and chemical energy, etc. by using batteries, capacitors, magnets, hydrogen, etc. The principal criteria of energy storage devices required for a specific application are to determine and maintain the reliability, durability, safety, and cost of the material. Moreover, the environmental impact, recyclability, and electric load of the material are also considered. The energy storage devices being used must have specific power and energy. The response and efficiency rates of the device must be adequate enough. The maintenance of the device must be simple so that the device can be handled easily. Moreover, the sensitivity of the device to heat must also be considered. The device must not have a higher discharge rate as it will cause the dissipation of the energy. The operating cost of the device must not be neglected [22].

5.1 Types of Energy Storage Devices

There has been meticulous research regarding clean energy storage materials in recent times [22]. Some of the energy storage devices are [23]:

- Battery Energy Storage.
- Mechanical Energy Storage.
- Thermal Energy Storage.
- Pumped Hydroelectric Energy Storage.
- Hydrogen Energy Storage.

5.1.1 Battery Energy Storage

Batteries are the most versatile and commonly used energy storage devices. These are the electrochemical devices that consist of cells having two terminals named cathode - positively charged and anode - negatively charged. The mostly used solid batteries are Li-ion and Pb acid batteries. Other solid battery types are Ni-Cd, Na-S, and Zinc-air batteries. Li-ion batteries have superior energy density compared to Pb-acid batteries. Liquid batteries are also being utilized as storage devices that include liquid electrolyte solutions such as; Vanadium redox, Fe-Cr, and Zn-Br. Capacitors are not characterized as batteries but are part of electrochemical technology and have applications at the sub-minute level response. Batteries are the ideal long-term storage devices since they can be used as backup and also useful for nighttime since they can be charged during the daytime [24,23].

5.1.2 Thermal Energy Storage

Thermal storage is the storage of high intensity or cold during the difference in conditions of storage medium for example from gas to fluid or solid to fluid as well as the other way around. Such sort of storage methods include the utilization of molten salt and liquid air. Molten salt has arisen as economically feasible with concentrated solar power however this and other heat stockpiling choices might be restricted by the requirement for huge underground storage caves [23].

5.1.3 Pumped Hydroelectric Energy Storage

Pumped hydro energy frameworks depend on enormous water supplies. Such energy frameworks require water cycling between two repositories at various levels with the energy stockpiling in the upper supply water, which is delivered when water is required to the lower repository. Whenever the power demand is high, generally at peak hours during the day, water is set free from the upper repository to the lower supply through a dam to produce power for the grid. At the point when power demand is low and there is an unnecessary measure of power accessible on the grid, the water is pumped back up to the raised repository. Pumped hydro is a drawn-out power age framework that backings. As indicated by Department of Energy data, pumped hydro energy storage gives 95% of current U.S. energy storage limit [24,23].

5.1.4 Mechanical Energy Storage

Mechanical energy storage includes the storage of kinetic forces of rotation and gravitation towards storing energy. These frameworks include the utilization of flywheels and compressed air frameworks. A flywheel is a fast wheel that pivots around a hub and stores energy precisely as kinetic energy. Basically, it is the same as the generators on the grid, the heavier the flywheel, the more energy it can store. The quicker the flywheel mass twists, the more energy there is to put away. Flywheels can be introduced in a vehicle or can act as a megawatt-scale energy storage framework. They can be immediately charged and discharged. They are utilized for the most part for stationary storage applications. The advantage of the flywheel is that it is a mechanical framework, making it an eco-accommodating arrangement, and an enduring gadget that offers numerous long periods of reliable performance. The Compressed Air Storage strategy compresses air into a cavern utilizing engines fueled by power or petroleum gas and when energy demand is high, the air is delivered through a turbine to create power. This energy stockpiling technique has been in need for quite a long time, particularly in the mining business [24,23].

5.1.5 Hydrogen Energy Storage

Energy storage with hydrogen includes the change of energy from electrical power through electrolysis for storage in tanks. This energy is additionally utilized for transport purposes and in the industry as a leftover. In this interaction, the electricity is changed into hydrogen and the converted electrical energy can be utilized as gas for fuel in a combustion engine or in a fuel cell. If a modest amount of electrical power is accessible, we can have hydrogen through the electrolysis of water, with generally high effectiveness. We can store the produced hydrogen depending on large-scale (in underground caverns) or short-scale (in steel compartments) applications. Hydrogen can be utilized as fuel for piston engines, or hydrogen fuel cells, gas turbines, and fuel cells offer the best effectiveness. Gas frames the reason for the hydrogen economy in which it replaces petroleum products in numerous combustion applications, and that's why hydrogen energy storage is of interest [25].

6.   Transport Properties

Transport properties are the molecular properties of materials that decide the rate of energy per unit volume moved by that material. These properties incorporate conductivity, diffusivity, viscosity, ionic exchange, and so forth. For assurance of intermolecular expected capacities and for the advancement of precise models of transport properties in dense liquid states it is essential to have exact information about transport properties to execute the ideal plan of the various things of chemical processing plants. For the comprehension of intermolecular forces, transport properties are of great interest. It is difficult to quantify these properties for all economically and hypothetically significant liquids; consequently, foreseeing these properties by utilization of appropriate hypothetical models is extremely fundamental. The model of gas transport properties has been sensibly explained by the use of the kinetic theory of gases; nonetheless, the model of liquid viscosity is ineffectively evolved [26]. As transport properties of molten salts, diffusion constants, and electrolytic conduction might be viewed as the principal properties, despite the fact that viscosity and heat conductivity are additionally included. Most molten salts are made of ionic species, or at least, cations and anions, which cooperate straightforwardly with one another principally by the coulombic attraction without solvents like water particles as intermediators. Among cations and anions, the coulombic repulsion works. Subsequently, molten salts might be viewed as the least complex objective of "columbic chemistry." In this sense, among the different properties of molten salts, quite possibly the most trademark one is electrical transport communicated as far as electric conductivities or mobilities [27]. Transport properties in GaAs/AlGaAs heterostructures, quantum wells, and silicon MOSFET structures have been investigated for a long time and are still under investigation. A gate can be utilized in order to adjust the electron density in MOSFET designs. Owing to interface-roughness scattering and impurities at the Si/Sio2 interface, the mobility in such designs is restricted. MOSFET designs have lower pinnacle mobility at low temperatures of about $\mu \approx 3 \times 10^4$ cm^2/Vs in comparison to GaAs/AlGaAs heterostructures. In GaAs/AlGaAs structures mobility is higher with a value of $\mu \approx 8 \times 10^6$ cm^2/Vs due to the remote doping [28]. In the field of solid-state material science, astronomy, and high energy-density plasma physical science, the transport properties of strongly coupled plasmas are of interest. Numerous hypothetical models have been introduced to depict transport phenomena in strongly coupled plasmas, but a couple of trials have been performed to contrast and estimations [29].

7.   Energy Density of Ferroelectrics

Properties such as high energy density and reliance on dielectric energy storage devices are vital in high pulsed power applications such as electrical-power devices and systems. In the field of batteries, the main focus has been on increasing energy storage density. Developments in comprehension of basic science and in the field of engineering have led to an enhanced storage density of generations of batteries ranging from lead acid to nickel-cadmium and nickel-magnesium batteries and to Li-ion batteries. Recent demands of ion battery life have caused urgency in the lofty energy storage density of batteries [30].

The recoverable energy stowing density (W_{reco}) is expressed by Eq.2 [6]:

$$W_{reco} = \int_{Pr}^{Pmax} E \, dP \tag{Eq.2}$$

and energy storage efficiency (η) can be stated by Eq.3,

$$\eta = \frac{Wreco}{Wreco + Wloss} \tag{Eq.3}$$

Where P_r stands for remnant and on the other hand P_{max} represents maximum polarizations respectively, and W_{loss} denotes the uncovered energy density instigated by the hysteresis loop.

7.1 Ways to Improve Energy Density

By enhancing the electrode's specific capacity or through increased cell voltage, we can boost the energy density of batteries. The optimization of electrode materials has optimized their voltage limit by virtue of the restricted capacity of batteries, owing to 10 years of improvement of the energy density of intercalation chemistry cells with a voltage of $\leq$ 4.4 V [31]. Storage density can be optimized by opting for alternative electrode materials.

7.1.1 Chemical Substitution

One of the most authentic methods used to enhance the energy density of ferroelectrics is the chemical substitution of energy storage devices. This method involves the substitution of external chemicals in devices to enhance energy density. The chemicals added to energy storage devices make ions as they find in the electrolytic solution. The ions thus generated increase the transport properties of the devices and hence the energy density of the devices (batteries etc.) increases. An experimental example regarding this process is mentioned below:

$(0.7-x)BiFeO_3-0.3BaTiO_3-xBaZn_{1/3}Nb_{2/3}O_3 + 0.1$ wt% MnO_2 abbreviated as (BFO-BTO-BZNO) where, x = 0.06, 0.10, 0.14 and 0.18, truncated as BZNO06, BZNO10, BZNO14, and BZNO18, individually ceramics were synthesized through an ordinary solid-state reaction process utilizing Bi_2O_3 (almost 100%), Fe_2O_3 (close to 100%), $BaCO_3$ (almost 100%), TiO_2 (close to 100%), ZnO (close to 100%), MnO_2(99%) and Nb_2O_3 (99.99%) powders as unrefined components. Taking into account that Bi_2O_3 is easily volatilizable at very high temperatures, a 1 wt % overabundance of Bi_2O_3 was introduced. First, the ball milling was implemented for 24 h, in order for the powders to blend completely and then calcination was performed at 1053-1073 K for 3 hours and the ball was processed again for 24 hours in C_2H_6O. After the second drying, pellet formation, followed by the addition of 5 wt % of the fastener PVB and then sintering at 1213-1233 K for 4 h was conducted. An X-ray diffractometer was used to distinguish the polycrystallinity. Hitachi S-3400N-II checking electron magnifying instrument (SEM) was implemented to study the morphologies of ceramics. The sintered ceramic was cleaned with many types of sandpapers in order to achieve 190 μm thick samples which were then sputtered with gold from both sides of surfaces in order to achieve an electrode with a width of 1mm to study -E circles and width of 11 mm for permittivity estimations. A standard ferroelectric test framework was implemented to study a sample drenched in silicon oil along with a 1 Hz driving field that is sinusoidal, for studying temperature-subordinate ferroelectric hysteresis. LCR meter was utilized in the frequency region of 1kHz-100 kHz and in the temperature region of 298-773 K with a warming rate of 2K per minute to obtain dielectric constant and loss tangent as a component of temperature and frequency [11,32].

8.   High Energy Efficiency Perovskite Solar Cells

Since the testimony of long-life solid PSCs in 2012, solar cells were grounded on Halogenated top Perovskite containing carbon-based cations such as Methyl Ammonium $CH_3NH_3PbI_3$ and Amidinium. HC (NH_2) $2PbI_3$ are larger due to their excellent photovoltaic power. Since understanding the basics of photo absorbers is directly linked to their photovoltaic execution, the optoelectronic behavior of organic Pb-based halide perovskites was studied to provide a vision into more efficient PSCs. Perovskite solar cell conversion efficiency has been established to be highly reliant on the superiority of the perovskite membrane [32].

Great interest in $CH_3NH_3PbI_3$ perovskite is not owed to its high efficiency, but due to the newly made formations. Metal halide-based devices procure output similar to traditional silicon solid-state dye-sensitized solar cells and organic or inorganic halides are deposited on the nanostructured titanium oxide layer by a one-step spin coating process. Optical measurements indicate charge transfer from perovskite to both TiO_2 (electrons) and HTMs (holes), the latter being faster. A record efficiency (h = 15.0%) was recently achieved by applying a sequential deposition process (originally developed by Mitzi and his colleagues to convert PbI_2 to $CH_3NH_3PbI_3$ in the pores of TiO_2).

Perovskite materials are being used vastly in solar cells and there have been developments in the field of organometallic halide PSCs in order to better comprehend the basic understanding of related mechanisms. The sequential process for manufacturing PSCs has an extra edge over current single deposition techniques in terms of efficiency and stability. There is a report indicating a PCE of 19.3% of a two-dimensional perovskite solar cell at room temperature with a basic mixture of interface engineering and a metal co-doped Spiro-OMeTAD [33].

9.   Ferroelectrics for Energy Storage Devices

In the field of energy harvesting, dielectric capacitors have been broadly considered on the grounds that their electrostatic storage limit is immense, and they can convey the storage energy in an exceptionally brief time frame. Brilliant significance for the efficient storage of electrical energy has been acquired by relaxer ferroelectrics-based dielectric capacitors. Energy storage of $BiFeO_3$-based relaxer ferroelectrics in mass ceramics production, multi-layers, and thin film structures have been greatly investigated. Relaxer ferroelectrics have a low dielectric loss, low remnant polarization, high saturation polarization, and high breakdown strength, which are the primary boundaries for energy storage. Energy storage in capacitors, ferroelectrics, anti- ferroelectrics, and relax or ferroelectrics are expected contenders for energy storage devices. Bulk ceramics, multi-layered, and thin films BiFeO3-based relaxor ferroelectrics performance as energy storage devices are continuously being deliberated [33].

Global warming represents a possible danger to the fate of the planet. Preceding ignition of petroleum derivatives builds the conjunction of carbon dioxide and other ozone diminishing substances in the Earth's environment, prompting hotter air and environmental changes. What's more, the exhaustion of petroleum derivative assets, main possibility for the energy market, represents a gamble of energy emergency in our current reality where the quantity of buyers is expanding step by step. Given the genuine danger to life on Earth and the dangers of the energy emergency presented by the utilization of petroleum derivative assets, a perfect energy progress is not a kidding thought. Nonetheless, environment-friendly power sources, for example, solar based, wind based, and geothermal are generally inconsistent. Subsequently, utilizing and storing sustainable power for future access is a troublesome errand. Power created from sustainable power sources offers gigantic chances to fulfill future energy needs and the achievability of clean energy transitions [34].

The energy storage apparatus of a dielectric relies upon its polarization under the use of an electric field. The dielectric under the applied electric field is polarized so an equivalent measure of positive and negative charges gather on the outer layer of the dielectric. All in all, an electric field that goes against the applied electric field is actuated inside the dielectric. The value of the instigated field increases dramatically after some time until its size approaches the greatness of the outer field. This cycle is called capacitor charging. In this manner, the incited electrostatic energy is stowed in the dielectric and can be utilized in the application when it is released by the load. How much energy (U) is put away can be acquired from the likely distinction between the dielectric voltage (V) and the charge (q) prompted in the terminals on the outer layer of the dielectric utilizing the accompanying condition. $U = Zq_{max}Vdq$ (where q_{max} is the most extreme measure of charge put away in the anode when the capacitor is completely energized, and dq is the measure of increment of charge obtained during charging [35].

9.1   Fuel Cells

Chemical power of hydrogen or different fuels is utilized to produce clean and effective electricity. Fuel cells can be specified by their storage capacities, as they are capable to use a variety of fuels. Fuel cells have great energy applications as they may be used for big structures like a software energy grid as well as minute structures in a computer. Fuel cells may be utilized in a huge variety of applications throughout a couple of sectors, which include transportation, industrial/ commercial/ residential buildings, and long-time period power garages for the grid in reversible structures. Fuel cells have numerous advantages over traditional combustion-based cells. Fuel cells have superior efficiencies than combustion engines and have power conversion efficiencies exceeding 60%. In comparison to combustion engines, fuel cells have decreased or zero emissions [36]. Hydrogen gas cells provide the most effective water, addressing crucial weather demanding situations as there aren't any carbon dioxide emissions. There is also no air pollution that creates smog and motive fitness issues on the factor of operation. As they have few shifting parts, fuel cells are being operated commercially. Fuel cells behave like batteries, yet they don't deplete and needn't bother with being re-energized. It produces power and heat for as long as it is filled with fuel. A power device comprises two electrodes, a cathode (positive) and anode (negative) that are sandwiched around an electrolyte. Fuel, for example, hydrogen is provided to the negative electrode, and the air is provided to the positive cathode. In hydrogen power devices, protons and electrons are produced from hydrogen by the catalyst at the anode, these charges separate ways to the cathode. Power is produced, as electrons go through the outer circuit. Heat and water are produced when protons relocate through the electrolyte to the cathode and are joined with oxygen [37].

Perovskite based Materials for Energy Storage Devices   Materials Research Forum LLC
Materials Research Foundations 151 (2023) 67-88   https://doi.org/10.21741/9781644902738-3

9.2   Photocatalysts

Photocatalysts are materials that absorb light and in turn excite the material to higher energy and provide this energy for a chemical reaction. Photocatalysts with adsorbents, like titanium oxide/graphite, titanium oxide/silicon dioxide, apatite, and titanium oxide /hydro have efficiency preferable to natural contaminations over solid monolithic photocatalysts. TiO_2 photocatalysts have magnificent photocatalytic properties; so, the disintegrated particles can't be assimilated. Accordingly, a low debasement rate of pollutants is possessed by TiO_2 photocatalysts. To beat this disservice, hybrid photocatalysts are important to deliver photocatalysts. An example of hybrid photocatalysts is the stacked nanoparticles of BiOBr on reduced $C_{140}H_{42}O_{20}$ for the photocatalyzed degradation of dyes. In electron movement and in declining the rate of electron-hole reconsolidation, reduced graphene plays a pivotal role which has a huge surface area, fantastic adsorption property along with lofty electrical conductivity, and in addition the storage ability and transport of electrons only when mixed with different constituents. Whenever noticeable light associates with the GBiOBr photocatalysts, positive holes are abandoned in the valence band (VB) when electrons are invigorated from the VB to the CB of BiOBr. In order to shape •OH extremists, holes that are present in VB of BiOBr are made to respond with the surface adsorbed H_2O and OH^-. The (•OH) assumes a basic part in the debasement of organic contaminants. A portion of the electrons that are generated by photocatalysts travel to the graphene surface from the BiOBr. This movement is indicated by the distinction in the fermi energy levels, i.e., from top to bottom fermi levels [38].

9.2.1   Characterization and Preparation of Photo Catalysts

The photocatalyst was readied from a precursor of a homogeneous combination of strontium nitride and Ta_2O_5, at a 193K N_2 climate, as follows. A predetermined measured amount of Sr metal was disintegrated in fluid NH_3 and Ta_2O_5 powder was suspended. At 298 K, fluid NH_3 gets disintegrated. The buildup was a homogeneous combination of Sr $(NH_2)_2$ and Ta_2O_5. The subsequent buildup was cleared at 673 K for 5 hours to break down Sr $(NH_2)_2$ in the blend into strontium nitride. The combination obtained by exhaust was viewed as a precursor of a photocatalyst. The photocatalyst was acquired by heat treating the precursor at a foreordained temperature for 10 hours in an N_2 climate. The crystal design of the photocatalyst was affirmed by XRD estimation, and the morphology was seen by SEM. Bright or visible diffuse spectroscopy was applied to quantify the light retention conduct of the obtained photocatalyst [39].

9.3   Capacitive Energy Storage Devices

In cutting-edge electronic gadgets and power frameworks, polymer composites with high dielectric constant and electrostatic energy stockpiling are generally utilized. Ferroelectric polymers are matrices of dielectric polymer composites because they have the uppermost dielectric constant of all known polymers. The dielectric constant and fracture strength of the polymer composite material must be simultaneously improved in order to have a high storage density. Different material structural layouts and processing systems have been created to expand the energy density of ferroelectric polymer composites [40].

Capacitors are devices that store electrical energy in the form of electric charges stored on the plate. When you connect a capacitor to a power source, the energy is released and when you disconnect the capacitor from the charging source, the energy is put away. In this respect, capacitors are similar to batteries. A few sorts of capacitors have been created and are monetarily accessible. Conventional dielectric and electrolytic capacitors store charge on equal conductive plates that have a generally little surface area, restricting their capacitance.

Regenerative capacitor memory is a sort of memory which utilizes the electrical qualities of capacitance to store pieces of information. As these memories ought to be recovered consistently (i.e., modify and peruse otherwise called revive) to save information loss. An electric field in a dielectric dipole can store energy by capacitors [41].

Conclusion

As the energy resources of the world are reducing rapidly, it is necessary to synthesize alternate resources such as energy storage devices, and ferroelectric perovskites which possess good energy storage ability, and exhibit strong polarization P_s with applied fields by changing the orientation of domains. Lead-based and lead-free ferroelectric perovskites have been discussed in this chapter. Lead-based perovskites have limited energy storage properties due to low dielectric breakdown strength. However, lead-free ferroelectrics provide environment-friendly energy resources. Both lead-based and lead-free ferroelectrics are discussed according to their energy storage properties and uses. Different types of energy storage devices are briefly discussed. Types of ferroelectric energy devices are fuel cells and capacitors. Ferroelectrics are widely used materials in the electronic world. However, their performance may vary from material to material. The performance of ferroelectrics can be enhanced by increasing their energy density which can be done by using chemical substitution as mentioned above.

References

[1] B. Saparov, D.B. Mitzi, Organic-inorganic perovskites: Structural versatility for functional materials design, Chem. Rev. 116 (2016) 4558-4596. https://doi.org/10.1021/acs.chemrev.5b00715

[2] M.K. Assadi, S. Bakhoda, R. Saidur, H. Hanaei, Recent progress in perovskite solar cells, Renew. Sust. Energy Rev. 81 (2018) 2812-2822. https://doi.org/10.1016/j.rser.2017.06.088

[3] P.R. Varma, Low-dimensional perovskites, in: S. Thomas, A. Thankappan (Eds.), Perovskite Photovoltaics, Elsevier, 2018, pp. 197-229. https://doi.org/10.1016/B978-0-12-812915-9.00007-1

[4] V.M. Goldschmidt, Die Gesetze der Krystallo Chemie, Naturwissenschaften. 14 (1926) 477-485. https://doi.org/10.1007/BF01507527

[5] A.R. Chakhmouradian, R.H. Mitchell, Compositional variation of perovskite-group minerals from the Khibina complex, Kola Peninsula, Russia, Canad Mineral. 36 (1998) 953-969.

[6] J.W. Anthony, R.A. Bideaux, K.W. Bladh, M.C. Nichols, Handbook of Mineralogy, Vol. 1. Mineral. Data. Publ., Cambridge, 2001, pp. 152-153.

[7] M.E. Lines, A.M. Glass, Principles and Applications of Ferroelectrics and Related Materials, Oxford university press., Britain, 2001, pp. 150-200. https://doi.org/10.1093/acprof:oso/9780198507789.003.0016

[8] L. Yang, X. Kong, F. Li, H. Hao, Z. Cheng, H. Liu, J.-F. Li, S. Zhang, Perovskite Lead-free dielectrics for energy storage applications, Progr. Mater. Sci. 102 (2019) 72-108. https://doi.org/10.1016/j.pmatsci.2018.12.005

[9] J. Valasek, Piezo-electric and allied phenomena in rochelle salt, Phys. Rev. 17 (1921) 475. https://doi.org/10.1103/PhysRev.17.475

[10] F. Bassani, Encyclopedia of Condensed Matter Physics, Elsevier acad. press., Amsterdam, 2005.

[11] P. Bhattacharya, R. Fornari, H. Kamimura, Comprehensive Semiconductor Science and Technology, Newnes., Oxford, 2011, pp. 77-200.

[12] A.P. Barranco, J.S. Guerra, Y.G. Abreu, I.C. dos Reis, Perovskite layer-structured ferroelectrics, in: B.D. Stojanovic (Eds.), Magnetic, Ferroelectric, and Multiferroic Metal Oxides, Elsevier, 2018, pp. 71-92. https://doi.org/10.1016/B978-0-12-811180-2.00004-9

Materials Research Forum LLC
https://doi.org/10.21741/9781644902738-3

[13] M.Y. Zhuravlev, R.F. Sabirianov, S. Jaswal, E.Y. Tsymbal, Giant electroresistance in ferroelectric tunnel junctions, Phys. Rev. Lett. 94 (2005) 246750- 246802. https://doi.org/10.1103/PhysRevLett.94.246802

[14] A. Kakekhani, S.I. Beigi, Ferroelectric-based catalysis: Switchable surface chemistry, ACS Catalysis. 5 (2015) 4537-4545. https://doi.org/10.1021/acscatal.5b00507

[15] L. Zhang, J. Miao, J. Li, Q. Li, Halide perovskite materials for energy storage applications, Adv. Funct. Mater. 30 (2020) 2003598- 2003653. https://doi.org/10.1002/adfm.202003598

[16] H. Zhang, T. Wei, Q. Zhang, W. Ma, P. Fan, D. Salamon, S.-T. Zhang, B. Nan, H. Tan, Z.-G. Ye, A review on the development of Lead-free ferroelectric energy-storage ceramics and multilayer capacitors, J. Mater. Chem: C, 8 (2020) 16648-16667. https://doi.org/10.1039/D0TC04381H

[17] H. Palneedi, M. Peddigari, A. Upadhyay, J.P. Silva, G.-T. Hwang, J. Ryu, Lead-based and Lead-free ferroelectric ceramic capacitors for electrical energy storage, in: D. Maurya, A. Pramanick, D. Viehland (Eds.), Ferroelectric Materials for Energy Harvesting and Storage, Elsevier, 2021, pp. 279-356. https://doi.org/10.1016/B978-0-08-102802-5.00009-1

[18] R.N. Perumal, V. Athikesavan, Investigations on electrical and energy storage behavior of PZN-PT, PMN-PT, PZN-PMN-PT piezoelectric solid solution, J. Mater. Sci. : Mater. Electron. 30 (2019) 902-913. https://doi.org/10.1007/s10854-018-0361-x

[19] E. Brown, C. Ma, J. Acharya, B. Ma, J. Wu, J. Li, Controlling dielectric and relaxor-ferroelectric properties for energy storage by tuning Pb0.92 La0.08Zr0.52Ti0.48O3 film thickness, ACS Appl. Mater. Interfaces 6 (2014) 22417-22422. https://doi.org/10.1021/am506247w

[20] J. Koruza, L.K. Venkataraman, B. Malič, Lead-free perovskite ferroelectrics, in: B.D. Stojanovic (Eds.), Magnetic, Ferroelectric, and Multiferroic Metal Oxides, Elsevier, 2018, pp. 51-69. https://doi.org/10.1016/B978-0-12-811180-2.00003-7

[21] L. Jin, F. Li, S. Zhang, Decoding the fingerprint of ferroelectric loops: Comprehension of the material properties and structures, in: L. Jin (Eds.), Progress in Advanced Dielectrics, World Scientific, 2020, pp. 21-104. https://doi.org/10.1142/9789811210433_0002

[22] R. Folkson, Alternative Fuels and Advanced Vehicle Technologies for Improved Environmental Performance: Towards Zero Carbon Transportation, second ed., Elsevier, 2014.

[23] I. Dincer, M.A. Rosen, Thermal Energy Storage Systems and Applications, third ed., John Wiley & Sons., Hoboken, 2021. https://doi.org/10.1016/B978-0-12-824372-5.00009-9

[24] Wind Power Engineering, An overview of 6 energy storage methods. https://www.windpowerengineering.com/an-overview-of-6-energy-storage-methods/

[25] P. Breeze, Power System Energy Storage Technologies, first ed., Academic Press, 2018. https://doi.org/10.1016/B978-0-12-812902-9.00008-0

[26] H.W. Xiang, The Corresponding-States Principle and its Practice: Thermodynamic, Transport and Surface Properties of Fluids, Elsevier, Amsterdam, 2005. https://doi.org/10.1016/B978-044452062-3/50005-1

[27] I. Okada, Ionic transport in molten salts, in: F. Lantelme, H. Groult (Eds.), Molten Salts Chemistry, Elsevier, 2013, pp. 79-100. https://doi.org/10.1016/B978-0-12-398538-5.00005-6

[28] A. Gold, Transport properties of Silicon/Silicon-Germanium (si/sige) nanostructures at low temperatures, in: Y. Shiraki, N. Usami (Eds.), Silicon-Germanium (SiGe) nanostructures, Elsevier, 2011, pp. 361-398. https://doi.org/10.1533/9780857091420.3.361

[29] R. Shepherd, D. Kania, L. Jones, D. Schneider, R. Stewart, The measurement of transport properties in strongly coupled plasmas, in: S. Ichimaru (Eds.), Strongly Coupled Plasma Physics, Elsevier, 1990, pp. 433-437. https://doi.org/10.1016/B978-0-444-88363-6.50059-5

[30] H. Tang, Y.-C. Hu, X.-Y. Chen, X.-D. Jian, X.-B. Zhao, Y.-B. Yao, T. Tao, B. Liang, X.-G. Tang, S.-G. Lu, Enhancement of energy-storage properties in BiFeO3-based Lead-free bulk ferroelectrics, Ceram. Inter. 48 (2022) 4-10. https://doi.org/10.1016/j.ceramint.2022.02.229

[31] L. Chen, X. Fan, E. Hu, X. Ji, J. Chen, S. Hou, T. Deng, J. Li, D. Su, X. Yang, Achieving high energy density through increasing the output voltage: A highly reversible 5.3 V battery, Chem. 5 (2019) 896-912. https://doi.org/10.1016/j.chempr.2019.02.003

[32] J.M. Ball, M.M. Lee, A. Hey, H.J. Snaith, Low-temperature processed meso-superstructured to thin-film perovskite solar cells, Energy. Environ. Sci. 6 (2013) 1739-1743. https://doi.org/10.1039/c3ee40810h

[33] W.S. Yang, J.H. Noh, N.J. Jeon, Y.C. Kim, S. Ryu, J. Seo, S.I. Seok, High-performance photovoltaic perovskite layers fabricated through intramolecular exchange, Science. 348 (2015)1234-1237. https://doi.org/10.1126/science.aaa9272

[34] X. Chang, W. Li, L. Zhu, H. Liu, H. Geng, S. Xiang, J. Liu, H. Chen, Carbon-based CsPbBr3 perovskite solar cells: All-ambient processes and high thermal stability, ACS Appl. Mater. Interfaces. 8 (2016) 33649-33655. https://doi.org/10.1021/acsami.6b11393

[35] A.F. Xu, R.T. Wang, L.W. Yang, N. Liu, Q. Chen, R. LaPierre, N.I. Goktas, G. Xu, Pyrrolidinium containing perovskites with thermal stability and water resistance for photovoltaics, J. Mater. Chem: C, 7 (2019) 11104-11108. https://doi.org/10.1039/C9TC02800E

[36] H. Zhu, K. Miyata, Y. Fu, J. Wang, P.P. Joshi, D. Niesner, K.W. Williams, S. Jin, X.-Y. Zhu, Screening in crystalline liquids protects energetic carriers in hybrid perovskites, Science 353 (2016) 1409-1413. https://doi.org/10.1126/science.aaf9570

[37] X.-Y. Zhu, V. Podzorov, Charge carriers in hybrid organic-inorganic Lead halide perovskites might be protected as large polarons, ACS Publ. 6 (2015) 4758-4761. https://doi.org/10.1021/acs.jpclett.5b02462

[38] K. Miyata, D. Meggiolaro, M.T. Trinh, P.P. Joshi, E. Mosconi, S.C. Jones, F.D. Angelis, X.-Y. Zhu, Large polarons in Lead halide perovskites, Sci. Adv. 3 (2017) 1701200-1701217. https://doi.org/10.1126/sciadv.1701217

[39] S. Wei, X. Xu, Boosting photocatalytic water oxidation reactions over Strontium Tantalum Oxynitride by structural laminations, Appl. Catal. B: Environ. 228 (2018) 10-18. https://doi.org/10.1016/j.apcatb.2018.01.071

[40] A.A. Ismail, D.W. Bahnemann, Mesoporous Titania photocatalysts: Preparation, characterization and reaction mechanisms, J. Mater. Chem. 21 (2011) 11686-11707. https://doi.org/10.1039/c1jm10407a

[41] M. Addamo, V. Augugliaro, A.D. Paola, E.G. López, V. Loddo, G. Marci, R. Molinari, L. Palmisano, M. Schiavello, Preparation, characterization, and photoactivity of polycrystalline nanostructured TiO2 catalysts, J. Phys. Chem: B, 108 (2004) 3303-3310. https://doi.org/10.1021/jp0312924

Perovskite based Materials for Energy Storage Devices
Materials Research Foundations 151 (2023) 89-110

Materials Research Forum LLC
https://doi.org/10.21741/9781644902738-4

Chapter 4

Techniques for Recycling and Recovery of Perovskites Solar Cells

Somi Joshi[1], Kanchan Chaudhary[1], Kalpana Lodhi[2], Manjeet Singh Goyat[3]and
Tejendra K. Gupta[1*]

[1]Amity Institute of Applied Sciences, Amity University, Sector-125, Noida 201313, India

[2]Advanced Materials and Device Metrology Division, CSIR-National Physical Laboratory, New Delhi, 110060, India

[3]Department of Applied Science, University of Petroleum & Energy Studies, Dehradun 248007, Uttarakhand, India

*tejendra.amu@gmail.com

Abstract

Perovskite-based photovoltaic cells (PSCs) have been viewed as a capable future generation candidate for photovoltaic innovation with a phenomenal improvement in power transformation productivity (PCE) because of various changes taking place over a decade due to the amazing optoelectronic qualities of perovskite materials. PSCs are extremely competitive in comparison to existing marketed silicon and thin film-based photovoltaic automation due to their configurable band gap, intense inclusion, high power transformation effectiveness, and minimal expense. Contrarily, commercial items invariably affect massive quantities of waste and the quality of products, both of which have a terrible impact on the environment. Perovskite solar cell recycling and recovery methods should be investigated and developed ahead of time to overcome this issue. As a result; PSC components must be recycled for industrial fabrication applications. The importance of recycling and several recycling approaches are discussed in this paper, along with the recycling of several parts of perovskite solar cells which include transparent conductive oxide (TCO) substrates, electron transport materials (ETM), metal electrodes, toxic lead material, and monolithic structure. The reusing technique has also been deliberated about the variety of layered designs. Lastly, the prospect of next-stage perovskite-photovoltaics reusing has been proposed as a means of encouraging eco-friendly extensive manufacture and use.

Keywords

Perovskite Solar Cells, Recycling Process, Lead Toxicity, Cost Analysis

Contents

Techniques for Recycling and Recovery of Perovskites Solar Cells **89**

1.   Introduction..**90**

    1.1   Recycling Roadmap...91

    1.2   Delamination of perovskite solar cell modules 93

3.   Need of recycling..**94**

    3.1   Degradation of perovskite solar cells 94

    3.2   Use of expensive raw materials...95

    3.3   Toxicity behavior of lead...96

4.   Recycling of several parts of perovskite solar cells **97**

    4.1   Recycling of transparent conducting oxide (TCO) 98

    4.2   Recycling of Electron Transport Layer (ETL).................................99

    4.3   Recycling of toxic lead component 100

    4.4   Recycling of metal electrodes..103

    4.5   Recycling of monolithic structure 103

5.   Future challenges...**104**

6.   Analysis of cost..**105**

Conclusion and future perspective...**106**

Conflict of interest...**107**

Acknowledgment...**107**

References **107**

1.   Introduction

Perovskite photovoltaic (PV) innovations are altering the power age by utilizing another age of metal halide perovskites (MHPs) [1]. Perovskite photovoltaic cells are viewed as an exceptionally encouraging contender for third-age sun-powered cells attributable to their near-perfect transparent construction, configurable direct band holes, huge ingestion coefficient, high ambipolar portability, long transporter dispersion length, little exciton restricting energy high imperfection resistance, arrangement processability, and low handling cost [2]. The majority of PSCs contain a poisonous element lead as well as other non-ecofriendly components, which will ultimately result in the leakage of toxic compounds, contamination, and material utilization, particularly due to the long-term presence within the living body and unknown repercussions. The expensive electrodes used in transparent conductive oxide (TCO) substrates are indium tin oxide (ITO), fluorine-doped tin oxide (FTO), and noble metal electrodes (such as gold), which cost around 15 and 20% of the overall cost of PSCs manufacturing. This significantly depletes rare resources, especially noble metals like gold and indium. Since the aforesaid materials are only used once, the manufacturing cost of PSCs skyrockets, making them prohibitively expensive to mass produce [3]. Because of the previously mentioned issues, we evaluated the advancement of PSCs' parts reusing, which includes TCO substrates with electron transport layer, TCO substrates (FTO and ITO), lead compounds (like lead iodide, PbI_2), metal anode furthermore, solid state device having transport layers. We have looked at the various possibilities for adopting the core form of PSCs and the creation of more eco-friendly adjustments and minimal expenditure as a result of their futuristic strength. At last, viewpoints of the following stage of PSCs reusing have been examined for minimal expense eco-friendly manufacturing, and expected modern application.

1.1 Recycling Roadmap

Items are normally landfilled at the end of life (EOL) in the conventional linear economy, and innovative products made from virgin resources are developed to substitute them. A circular economy, on the other hand, optimizes power and material consumption over product life cycles by remanufacturing, renovating, revamping, reprocessing, and recycling EOL products, as well as expanding initial product lifetime to raise the quantity of valuable facility per product; when components cannot be re-manufactured, renovated, mended or reused, reutilizing is the final geometric choice. Various items, however, made up of complicated mixed elements are difficult to dispose of [4]. When items are taken from service, design for recyclability (DFR) emphasizes making compounds that can be securely and inexpensively reprocessed using current reutilizing technology and processes, as well as changing materials to make products that are safer for the environment and humans (**figure 1**). Replacement of particular harmful or uncommon components may not be possible in some instances. In these situations, collaboration among producers and recyclers is critical for assuring reprocessing and controlling the ecological concerns that follow. The following are the conditions that should be taken care of while recycling any compound [5].

- Functionality, durability, resilience, dependability, and affordability are essential product needs; DFR should enhance these features but may result in trade-offs between recyclability and product performance and cost.

- DFR outcomes are dependent on material selection and the capacity to free separate materials.

- By reducing unsafe materials in goods or improving these materials fully via DFR, recycling outcomes can be improved.

- Reducing and controlling difficult-to-recycle materials can help to increase total recycling yield.

- Disassembly and material release can be aided by reducing irreversible adhesives or comparable bindings, specifically over full surfaces for different materials.

- Recyclability can be improved by design for disassembly (DFD).

- Estimating recyclability gains and economic or environmental impacts as a result of DFR is critical for ongoing improvement, trade-off identification, and value communication.

- Recyclers can classify feedstocks more easily by using tags to recognize recyclable and non-recyclable items; labeling consistency is critical for uptake and usage.

- Circular manufacturing is aided by product designs that include recycled materials.

- Long-term identification of the construction of the compound and arrangement could make recycling operations harmless and more effective.

- The composition of the back sheet has a significant impact on recyclability.

- Metal selection has a big influence on recycling processes and costs.

- PV module disassembly can be made easier by reducing encapsulant use or employing reversible encapsulants.

- Recyclability and economics trade-offs arise when the number and complexity of module materials are reduced.

- Using various sealants in the aluminum frame may allow component separation without causing module damage.

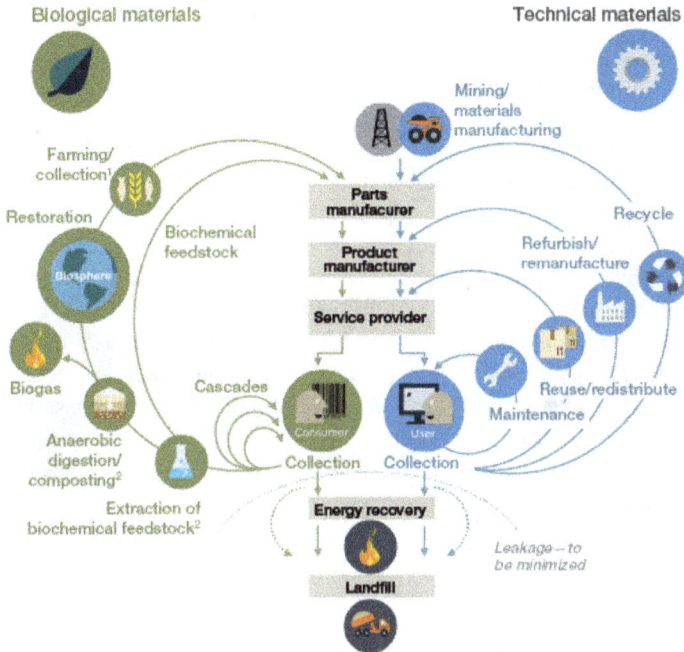

Figure 1. The role of design for recyclability. Reprinted under an open-access Creative Common CC BY license [6].

1.2 Delamination of perovskite solar cell modules

Delamination is a significant topic for tandem configuration perovskite cell reliability, especially when it comes to module manufacturing. The exemplification interaction makes the stressors that trigger the delamination because of the variety in the extension coefficients of the films utilized in the cell. The first stage in recycling EOL perovskite solar modules is to create a delamination technology for disassembling the encased modules and exposing the perovskite layer. The perovskite module structure is built on ITO glass which is further enclosed with another glass piece and the encapsulant in this study. Epoxy resin, polyolefin, surlyn, polyisobutylene, polyurethane, and other encapsulants for perovskite solar cells are already available. These encapsulants together with a back cover glass successfully prevent the passage of moisture, oxygen, and other risks [7]. This was chosen for this investigation because it is the extremely liable configuration in marketable perovskite solar panels that provides the best stability. We observed that a brief high-temperature thermal treatment may efficiently disintegrate encapsulated perovskite solar modules, resulting in undamaged back cover glasses and ITO. Following 2 minutes of thermal stress at 250°C, the melting of the polymer encapsulant takes place, which puts the perovskite solar module under strain between the interface of the electron-transport layer (ETL) and metal electrode, in which epoxy resin is used as an encapsulant. Using 1,2-dichlorobenzene, the electron-transport layer was washed. For further lead recycling, DMF was used to dissolve the lead halide perovskite. ITO/glass may be reused for module refabrication, only after a proper wash of the hole-transport layer (HTL) and other leftovers. After reusing the ITO/glass substrate, no discernible conductivity changes were observed. The conductivity of ITO/glass was marginally improved from 14.6Ω/sq to 15.2Ω/sq after annealing at 250°C for 1 hour. The copper and chromium electrodes with the epoxy encapsulant were still present on the back cover-glass side after thermal stress, where a 30 nm Cr layer on the top surface produced a black hue and the electrode/encapsulant film produced wrinkles [8]. We scrape the metal electrode and encapsulant using a knife, however, the encapsulant remains still soft, and the rear cover glass became clean and ready for use.

3.   Need of recycling

For both technological and environmental objectives, as well as international legislation, recuperating precious substances and reprocessing hazardous metals from PSC products are essential. Although PV product recycling and recovery technologies were rapidly improved, PSC recycling and recovery technologies are still in their infancy. Several organizations have proposed studies to address this problem. The proposed technologies are categorized according to the PSC element they influence:

I.   Transparent conducting oxide (FTO/ITO)

II.   ETL

III. Perovskite layer (MaPbI$_3$)

IV. HTL

V.   Back electrodes.

3.1   Degradation of perovskite solar cells

As a result of the natural synthesis of perovskite materials and the responsiveness of interior outer variables, like stage conduct, crystallinity, ionic relocation (like iodide and lithium particles), interface properties, electrical properties, optical factors, reaction with oxygen, temperature, and humidity, are basic difficulties, which should be tended to for further developing stability, hybrid (organic/inorganic) perovskite solar cells (PSCs) are generally unstable [9]. Furthermore, moisture degradation occurs often and commonly causes reversible and irreversible damage to perovskite films, particularly irreversible deterioration. This deterioration procedure has to be completed in a matter of hours until it was unencapsulated, and it will be expedited by the addition of light irradiation and heat stress.

The degradation of MAPbI_3-perovskite is typically distributed into four stages: the creation of methylammonium iodide (MAI) and PbI_2 (Equation 1) in the presence of H_2O, decomposition of MAI into CH_3NH_2 and hydroiodic acid (HI) (equation 2), and HI continuous breaking into H_2 and I_2 in the presence of oxygen and light illumination (equations 3 and 4) [10].

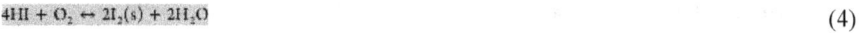

$$CH_3NH_3PbI_3 \xrightarrow{H_2O} CH_3NH_3I(aq)+PbI_2 \tag{1}$$

$$CH_3NH_3I(aq) \leftrightarrow CH_3NH_2(aq)+HI(aq) \tag{2}$$

$$2HI(aq) \leftrightarrow H_2 \uparrow +I_2(s) \tag{3}$$

$$4HI + O_2 \leftrightarrow 2I_2(s) + 2H_2O \tag{4}$$

Over time, the perovskite materials' semiconductor characteristics deteriorate the performance of PCE, PSCs, and operating stability. Additionally, the stability of PSCs has reduced due to the carrier transport materials (CTMs) oxidation, because, CTMs contain a minimum of one of the organic substances, polymers, and small molecules, which are vulnerable because of oxidation in ground states as well as in excited states. The deterioration of CTMs would cause perovskite decomposition by speeding up moisture and oxygen penetration [11].

3.2   Use of expensive raw materials

The cost of TCO substrates and metal electrodes contributes to the high-cost ratio of PSC production. According to prior reports, the costs of hole transport materials (HTM), gold (Au), indium tin oxide (ITO), and fluorine-doped tin oxide (FTO)are approximately 40, 25, 2000, and 100 dollars per square. Furthermore, because Au and indium are only used once, they will consume noble elements and waste rare resources, making PSCs less competitive in the photovoltaics market[12]. The cost of various layers in PSCs is shown in the pie chart (**figure 2**).

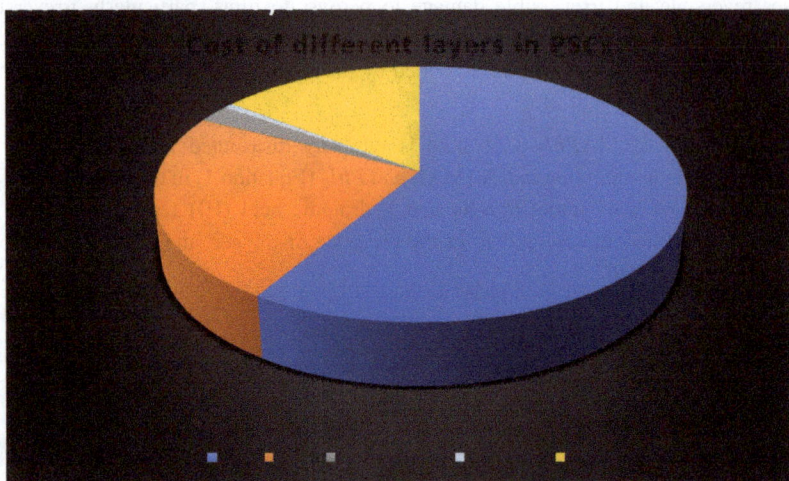

Figure 2. Pie chart showing the fabrication cost of various types of films in PSCs.

3.3   Toxicity behavior of lead

As shown in **figure 3**, Pb is commonly thought to be the main cause of toxicity in PSCs, prompting numerous LCA (life cycle assessment) investigations. Though, Sn is a key component as a Pb substitute in perovskite layers, which is linked to negative ecological and health impacts. The highest limit for acceptable lead levels in the blood for adults is 10 mg/dl and for children, the value is decided at 5 mg/dl, although the typical range is 2-9 mg/L for tin levels in human blood. The toxic behavior of Pb and Sn must be studied and taken into account for the commercialization and industrialization of PSCs. Pb has been found to pass into the human body via the gastrointestinal, respiratory, and cutaneous pathways, producing enzyme and receptor dysfunction and toxic heavy metals in breastfed newborns. The chronic revealing of tin was discovered to have a comparable influence to Pb poisoning and can cause substantial injury to the human body [13].

PSC toxicity is a severe and avoidable hazard, hence several LCA studies have been conducted to determine the actual ecological effects. Pb leaching from PSC panels over time is the primary source of concern for PSC toxicity. When compared to background Pb levels in metropolitan areas, Pb from the leachate of a broken PSC panel was found to cause a low degree of pollution, which is undesirable but not catastrophic for the environment. The overall quantity of Pb utilized in PSCs would still be substantially lesser than the emissions radiate from cars and airplanes each year if PSCs were employed to meet the United States' need for power. The toxicity capacity of PSCs was recently determined to be around 20 times lesser as compared to that of grid energy, according to Baxter's research[14]. In comparison to other Pb-containing items, the toxicity from the lead-based solar cell is not a severe problem as compared with its lifetime, as revealed by many studies. However, when commercial PSC products are widely used, after the completion of the PSCs lifetime, a huge amount of PV waste will remain a key concern, hence PSC recovery and recycling techniques should be thoroughly established.

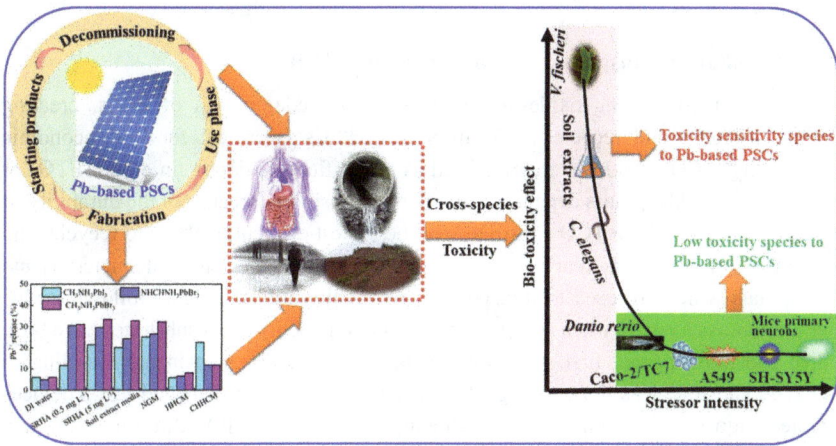

Figure 3. Toxic behavior of lead associated in perovskite solar cells. Reprinted with permission from [15].

4.   Recycling of several parts of perovskite solar cells

For both the technological and global objectives as well as international legislation, retrieving precious minerals and recycling hazardous metals from PSC is essential. Directive 2012/19/EU was adopted by the European Union in 2012 to handle and control waste electrical and electronic equipment [16]. Producers must assume accountability for garbage collection, recycling, and recovery under the legislation. Furthermore, China being one of the world's major markets has similar laws. As a result, these guidelines will control the commercialization of PSCs, emphasizing the necessity of perovskite recycling and recovery technologies. Although solar product recycling and recovery technologies are

Perovskite based Materials for Energy Storage Devices   Materials Research Forum LLC
Materials Research Foundations 151 (2023) 89-110   https://doi.org/10.21741/9781644902738-4

rapidly evolving, PSC recycling methods are still in their early stages of development. With their proposed studies, several groups have attempted to overcome this problem. In this section we are going to deal with the recycling of several parts of PSCs which are listed below:

- Transport conducting oxide (TCO)
- Electron transport layer (ETL)
- Toxic lead component
- Metal electrode
- Monolithic structure

4.1 Recycling of transparent conducting oxide (TCO)

As transparent conducting oxides (TCOs) are costly components of PSCs, creating solutions to reuse and recycle TCOs from trash PSCs is critical for PSC economic sustainability. TCO substrates are classified as either fluorine-doped tin oxide (FTO) or indium tin oxide (ITO) glass depending on their composition. Because of the durability of the oxide materials, these TCO substrates are the easiest element of PSCs to recycle, and they don't require any particular treatment other than washing the perovskite materials and the materials which are used to transport carriers with various solvents following PSC deterioration. As a result, when compared to other components, TCO substrates have been studied and implemented first. As shown in **figure 4**, the first step in the recycling of FTO/glass substrate was the cleaning of other components from damaged PSCs devices using techniques such as immersion in dimethylformamide (DMF), ultrasonication, and solvent washing. The solvents used for rinsing are chlorobenzene, acetone, ethanol, and deionized water [17]. The FTO/glass collected substrate is maintained with a similar composition, transmittance, and sheet resistance when compared to fresh. The solar cell structures consist $CH_3NH_3PbI_{3X}Cl_X$-of perovskite, hole transport layer (spiro-OMeTAD as HTL), on FTO/glass substrate with a silver electrode was deposited, and the power conversion efficacy (PCE) was 9.97%.

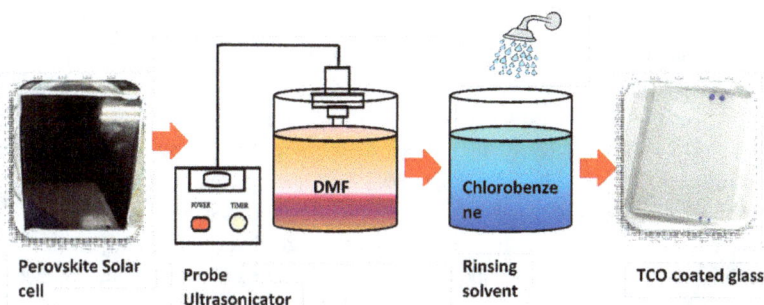

Figure 4. Schematic of the chemical route for the recycling of transparent conducting oxide (TCO).

After the extraction of FTO/glass, the performance of PSCs with ETL-free is yet insufficient, especially given the considerable fall in PCE. As a result, Binek et al. [12] were able to recycle TCO substrates whereas TiO_2 was used as an ETL. The compact TiO_2 thin film was grown via sol-gel technique in which ~0.40 mM titanium isopropoxide solution and a ~27 mM HCl solution in 2-propanol were used as a liquid precursor. Water, C_6H_5Cl, sticky tape, and DMF were used to wash or remove the components of MAPbI_3-perovskite, spiro-OMeTAD, Au, and TiO_2. ITO was recycled from degraded PSCs and patterned ITO/glass substrate was recycled by Augustine et al. [18] using KOH solution to dissolve aluminum electrode, PCBM, calcium buffer layer, poly-3,4-ethylenedioxythiophene, and MAPbI3-perovskite. Synced thermography was utilized to assess the electrical uniformity of sample surfaces. To measure the electrical homogeneity of the samples' surfaces, synchronized thermography was used. In comparison to the reference ITO, the ITO substrate had homogeneous electrical characteristics when treated with varying concentrations of KOH. Except for the 1.5 M, the KOH solution with a low concentration altered dramatically the electrical uniformity of the extracted ITO substrate. In this recycling, the photovoltaic functioning of PSCs reduced from 8.15 to 7.20 percent with a 1.5 M concentration of KOH. Even though the PSCs could maintain the photovoltaic performance that was essentially identical to that of new TCO/glass substrates, but were not cost-effective and free from pollutants, particularly the waste of ETL components and metal electrodes.

4.2 Recycling of Electron Transport Layer (ETL)

The TCO substrate with ETL film has been utilized to enhance PSC reusing productivity and improve reuse duration. Kim et al. [19] used polar-aprotic solvents such as dimethylformamide, dimethyl sulfoxide, or butyrolactone to eliminate the MAPbI$_3$-perovskite film and spiro-OMeTAD HTL, resulting in the regeneration of the mesoporous TiO_2 layered transparent layer with FTO/glass substrate after annealing it at 500°C for 1 hour. After looking at the figure given below, we can say that apart from a minor drop in fill factor (FF) and open-circuit voltage (V_{OC}) due to the surface morphological modification of TiO_2 after recycling, the PSCs can maintain a good photovoltaic performance after reprocessing ten times. After reprocessing the TiO_2/FTO/glass substance, Kadro et al. [20] achieved a relatively high PCE by utilizing an accelerated dismantling procedure with chlorobenzene and DMF solvents for the metal contact. After second-time recycling, the $MAPbI_3$ perovskite, HTL layer as spiro-OMeTAD, and Ag/ Au electrode were deposited on the TiO_2/FTO/glass substrate, resulting in nearly steady photovoltaic performance (16.1 % perovskite efficacy) in comparison to PSCs on a fresh substrate of TiO_2/FTO/glass. In the recycling of the TiO$_2$/FTO glass substrates in planar as well as mesoporous form, Huang et al. [21] used a method that included DMF and ultrasonication of chlorobenzene, DMF drenching, and treatment with ethanol, acetone, and then deionized water, after that ultraviolet light exposer. This prevented the spillage of lead content and environmental contamination. All the above-mentioned methodologies require high temperatures and complex procedures which lead to extra cost hence to reduce extra cost, Zhao proposed a new method in which he used the approach of chemical bath deposition for producing ZnO nanorods as ETL on aluminum doped zinc oxide. These rods can be further recycled by using DMF for washing perovskite [21].

4.3   Recycling of toxic lead component

Besides recycling the substrates of ETL/TCO/glass, the utmost essential lead component as PbI_2 is a key problem of Pb-PSCs since the Pb concentration raises worries about the possible venomousness of perovskites and their effects on biotic systems and the surroundings. Furthermore, due to solubility and purification challenges, the Pb content is the utmost challenging element to isolate and recycle. As a result, Pb content recycling has been both beneficial and challenging problem. Although, the recovery of lead from the PSCs is important due to toxicity issues even though it is not very expansive that requires economic recovery from waste PSCs. The first technique that is used to recycle lead was an electrochemical method, developed in 2016. The perovskite layers in PSCs with glass/TiO_2/FTO/$MAPbI_3$, $FAPbI_3$ architecture was dissolved using a solvent consisting of choline chloride and ethylene glycol. Using electrodeposition in a perovskite-containing solution, on a Pb electrode, Pb was applied and removed, as shown in **figure 5**. In this investigation, up to 99.8% of Pb was recovered from the solution [22]. In addition, a glass/FTO/TiO_2/$MAPbI_3$/spiro-OMeTAD/Au architecture was designed to reprocess Pb and use again FTO from PSCs. By removing PSCs layer by layer, this method was successful in keeping contaminants out of the recyclable materials. The perovskite film on the PSC was converted into MAI and PbI_2 in water when the Au electrode and HTL were removed, and the MAI was subsequently recovered by water. DMF solvent was utilized to dissolve and recycle the PbI_2.The recycled PbI_2 and FTO substrate were then utilized to make new PSCs. The efficiency of PSCs produced using this methodology was 13.5% which is somewhat less than PSCs made up of pure PbI_2 i.e., 14%. A minor quantity of contaminants in the recycled PbI_2 is most likely to blame for the reduced efficacy. Pb cyclic utilization in carbon-based PSCs was proven in subsequent research. $Pb(OH)_2$ was obtained from Pb in PSCs based on carbon with a glass/FTO/c-TiO_2/m-TiO_2/$MAPbI_3$/carbon structural design by simply treating them with $NH_3.H_2O$. PbI_2 was recovered after hydroiodic acid treatment, with a lead recovery rate of 96.7 percent. The efficiency of refabricated carbon-based perovskite was 11.36% while it was 12.17% for perovskite photovoltaic cells prepared by commercial PbI$_2$, [23].

DMF was initially used to dissolve MHP and delaminated perovskite solar modules to recycle the toxic lead. To eliminate lead ions from DMF, a carboxylic acid-based cation-exchange resin was employed. The lead ions sorbed on resin were eluted into an aqueous solution using HNO$_3$ as an eluent. When NaI was added to a solution containing Pb(NO$_3$)$_2$, PbI$_2$ gets precipitated, allowing recyclable components to be used to refurbish modules. Lead in the perovskite layer disintegrates naturally during the delamination of embedded perovskite solar-based modules. To completely remove lead from the natural dissolvable systems, lead particles were first adsorbed by a lead adsorbent. They were then transferred to the clean dissolvable and precipitated into PbI$_2$ for reuse. To reuse lead in retired perovskite solar-based modules, we chose carboxylic corrosive cation-trade pitch. The cation exchange between H$^+$ and Pb^{2+} depends on the lead-adsorption interaction and the lead-discharge process on saps.

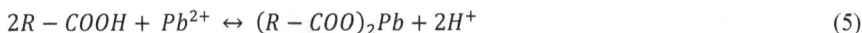

$$2R - COOH + Pb^{2+} \leftrightarrow (R - COO)_2Pb + 2H^+ \tag{5}$$

Reversibility of the ion exchange process should be observed. The equilibrium (in equation 5) can be reversed by a large concentration of H+ ions. It involves the regeneration of resin and lead is released. Cation-exchange resins frequently use acid solutions with high concentrations of acids such as H$_2$SO$_4$ or HCl. Because lead separation is challenging due to the low solubility of PbCl$_2$ and PbSO$_4$ in aqueous solution, regeneration with H$_2$SO$_4$ and HCl might precipitate directly the liberated Pb^{2+}ions as PbCl$_2$ and PbSO$_4$, respectively. During regeneration, the adsorbed lead ions are released in the form of water-soluble Pb(NO$_3$)$_2$ using a solution of HNO$_3$ aqueous. The PbI$_2$ is the best form of recycled lead, which serves as the primary lead supply for the majority of highly effective perovskite solar cells. Since, it is the primary source of lead for the majority of extremely effective perovskite solar cells, PbI2 recycled lead is the best kind for reuse. By adding inexpensive NaI, lead is changed from Pb(NO3)2 to PbI2 precipitate due to the difference in solubility in an aqueous solution. It is important to take Pb and Sn toxicity into account before commercializing and industrializing PSCs. Lead enters the human body via the skin, respiratory, and gastrointestinal systems. Prolonged contact with Sn is effective as compared to the poisoning of Pb, damaging the human body. These methods have attempted to solve the Pb problem in PSCs [24]. Although these tests had good lead recovery rates and the efficiency of these PSCs manufactured through recycled PbI_2 were marginally lower than those made with fresh PbI_2. Contaminants in recycled PbI_2 have been discovered as a possible source of performance degradation. To achieve fabrication requirements, raw materials which are utilized in electrical and electronic industries are frequently required to be extremely pure. To prepare for huge quantities of PSC devices, purifying procedures for recovered materials, as well as recycling and retrieval procedures, should be enhanced. Furthermore, because PSCs with more intricate structures have superior power conversion efficiency and stabilities, separating and recovering extremely pure Pb-containing source constituents from PSCs would be difficult.

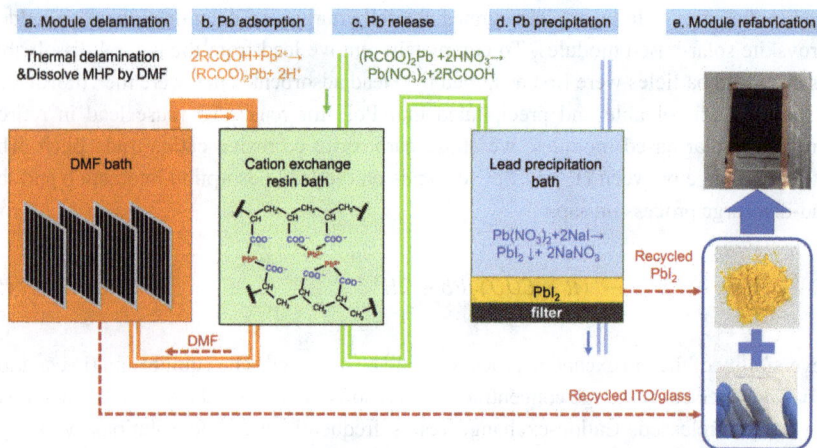

Figure 5. Recycling of toxic lead compound from Pb-based perovskite solar cells. Reprinted under an open-access Creative Common CC BY license [24].

4.4   Recycling of metal electrodes

Metal compounds, particularly the Au electrode, are commonly employed as contact electrodes for high photovoltaic efficiency PSCs. Because noble Au materials are costly, the Au electrode has added 20% to the price value of PSCs. As a result, the electrode of Au recycling is crucial for minimizing noble material use, manufacturing costs, and pollution. After eliminating the perovskite layer and HTL, a polar aprotic solvent such as DMF, GBL, or DMSO might be used to recycle the electrode of Au. However, due to the lightweight, tiny particles, and the impossibility of direct transfer, it is difficult to grow directly the electrode from the recycled gold particles via a vacuum-based thermal evaporation technique. Previously, when scientists recycled the Au electrode, they were unable to immediately re-use it due to its low mass and inability to survive against the suction created during pumping. Some organic HTM was detected in the Au electrode coating and thus needed to be cleaned with HCl. Hence, we need to develop a new approach for recycling metal electrodes. F. Yang et al. established a mechanically dried transfer procedure to load nanoporous Au films because of the high surface area for PSCs contact electrode, Au film is placed on a supporting membrane film [25]. When compared to PSCs created using a multi-step processing methodology, these cathode-coupled perovskite solar cells have achieved the maximum efficiency of 19% during the reverse check. For the recycling of the nano-Au electrode, the PSCs were dissolved in acetone and then dried at room temperature under standard conditions. The efficiency of the regenerated Au film was 16.45%, significantly higher than that of the evaporated Au layer. Application of this recycled nano-Au can be used in a flexible substrate to improve yield resistance. As per the calculations of the entire device cost, this unique mechanical transferring of approach nano-Au film might cut power source and precious metals consumption, lowering environmental hazards [26]. Furthermore, this approach might overcome the difficulties of traditional high-temperature and vacuum deposition processes, and it could be used in flexible PSCs.

4.5   Recycling of monolithic structure

For streamlining the PSC recycling procedure, ETL, HTL, contact electrode, and TCO-coated glass substrate have all grown to be well-liked options. Mesoporous NiO was used as the back electrode after printing an Al_2O_3 spacer layer of 1 mm thickness over the TiO_2 ETL surface. To make PSCs devices, the perovskite was then coated on an m-NiO/Al_2O_3/m-TiO_2/c-TiO_2/FTO/glass substrate. After dissolving the degraded perovskite in DMF and ethanol and drying it on a hot plate at 100°C for 10 minutes, the perovskite was reloaded on the aforementioned substrate. The efficacy of recycled monolithic substrate and PSC performance dropped from 13.6 to 12.1 percent, with JSC and FF performance dropping the most [27]. On testing ten separate devices, this recycled method provides a good level of reproducibility. According to scientists, the drop in photovoltaic efficiency after recycling is due to some degradation of the m-NiO layer during the washing process.

5.   Future challenges

PSCs devices rely heavily on absorber perovskite films, however, their instability and Pb carcinogenicity pose problems for PSCs waste management. Recycling and recovering perovskite films from trash PSCs devices is one viable option. The thermal efficiency deterioration of planar PSCs along with glass/FTO/TiO_2/$MAPbI_3$/spiro-OMeTAD/Au design was reported in 2018 and it was suggested that the distortion of spiro-OMeTAD HTL at high temperatures might be implicated. As a result, untouched perovskite films could be reprocessed. The thermal deformation of spiro-OMeTAD HTL takes place, which was eliminated using chlorobenzene but the perovskite layer film from the thermally damaged PSCs stayed on the FTO/TiO_2 substrate. A fresh spiro/OMeTAD layer was deposited on the recovered system. X-ray diffraction (XRD) and scanning electron microscopy (SEM) were used to examine the structural variations between the perovskite films during heating. The findings revealed that during heating, the perovskite films had identical structures. Furthermore, MAI is spin coated in isopropyl alcohol and after that annealed, finally, fresh perovskite film is recovered from the residual and PbI_2 left behind on damaged PSCs. In a study conducted, a single-step dissolving process with acetate and chloride and order deposition techniques for recycling perovskite layers were compared. The single-step acetate method generated perovskite films more efficiently as evidenced by photoluminescence efficiency and crystal structure analyses. These findings show that recycled perovskite films can significantly preserve their photoluminescence efficacy while maintaining their crystal structure. A thermal technique was used to create recycled PSCs with an increased PCE of 14.84%.

The difficulties of reprocessing HTL and the feasibility of perovskite film recycling from PSCs are both highlighted in this research. In the case of organic spiro-OMeTAD HTL functioning in the recycled cells, it was challenging to sustain as spiro-OMeTAD may easily deform because of crystallization, diffusion of Au, and photo-oxidation. Perovskite-recovery investigations must be carried out with a particular focus on scaling to vast areas. PSCs may have fewer significant environmental implications than the current PV products if the perovskite layer is recovered and reused on larger substrates. Furthermore, because present perovskite and HTL recycling methods are limited to $MAPbI_3$ and spiro-OMeTAD, recycling strategies for other perovskite films and HTL should be researched.

6. Analysis of cost

A techno-economic evaluation was completed to figure out the expected expense of recycling the perovskite modules. In this, we are mainly concerned with the cost of materials. The whole cost of material for a perovskite solar module depending on architecture is predicted to be $24.8/m^2$, which was nearly identical to the piece modeling by Li et al. [28] and Cai et al. [29]. The entire cost of recycled components was rough $12/m^2$, comprising front ITO/glass, PbI_2, and back cover glass[28]. Because the perovskite raw material accounts for such a low fraction of the total cost of material in perovskite solar modules, the price savings from the recycled lead as PbI_2 is not significant. The price of ITO glass and cover glass account for the most of the material cost. Compounds like DMF, cation-exchange resin, DCB, HNO_3 are consumed in recycling technique while we can still use some of these compounds in multiple cycles.

For making 1 mm^2 perovskite thin film, 4 g of DCB, 63 g of DMF, 20 g of resin, 2.5 g of HNO_3 and 2.7 g of NaI were used. The material used for recycling the perovskite solar module costs $4.24/m^2$ for this recycling technique assuming the use of components was only once. The expense of material can additionally be diminished to $1.35/m^2$ if reusing DMF and resin multiple times [30]. Aside from removing harmful lead from perovskite photovoltaic cells to protect the environment, this recycling technology generates significant cash, making recycling appealing. When compared to the manufacturing of new materials, recycled elements could save energy, provide another source of raw materials which do not reliant on primary mining, and alleviate several supply chain limitations.

Name of component	Cost ($/Kg)	Cost ($/m^2$)	References
ITO	1.28	6.4	[30]
Back glass	0.48	2.4-5.04	[29]
DMF (reusable)	38	2.41	[30]
Resin	60	1.20	[31]
DCB	22	0.09	[29]
HNO_3	18.6	0.05	[32]
PbI_2	1028	3.12	[32]
FAI	1480	1.55	[30]

Conclusion and future perspective

The low-cost and high PCEs represent the PSCs as a new and quickly expanding solar technology with the potential to govern the market of photovoltaic industries. End-of-life PSC items in big quantities will have a significant impact on the environment. PSC recycling and recovery technologies are currently available, as well as information on how to get them to create a full PSC-recycling process consideration. Glass/TCO substrates are the most expensive parts of PSCs (even TiO_2 ETLs) and can be used to make new PSCs have better efficiency, reducing the need for critical evaluation equipment (e.g., In, Sn). Regrettably, ongoing research provides little data about the reuse of PSCS, because perovskite films and HTLs sold commercially may have several elements that have yet to be investigated for recyclability. Because of the increasingly intricate configurations in the layers of perovskite in PSCs, Pb recycling and recovery processes have to be updated and refined. PSCs can be recycled in two ways: layer-by-layer or in a single phase, and while both approaches have been demonstrated, more research and optimization are needed. Even though it has a maximum PCE of 25.5% for a single junction and 29.1% with a tandem structure of perovskite/silicon but the unstable behavior of perovskite materials still poses a challenge to further research and industrial implementation due to long-term energy payback. Furthermore, the production and degradation of lead components in perovskite materials frequently create environmental problems due to the possibility of lead leakage. Furthermore, there is an increment in the cost of noble metallic materials and the number of noble materials used for single use. As a result, the recycling of PSC components has become an increasingly important research issue for environmentally responsible production and use.

Although a lot of efforts evolved into the recycling process, still, acquirement of recovered PSCs is insufficient due to low photovoltaic action for higher cycles of recycling. Therefore, even if the problem of perovskite photovoltaics long-term stability is eventually resolved, it is required to devote additional effort into recycling research. A future roadmap for recycling and recovering PSCs can be created based on the summary remarks above from current technology. Every significant PSC component must be recovered to raw materials, except for glass/TCO substrates, which may be reused as components. For the highly competitive market environment, the standards and criteria for the electronic product business are continuously modified, and EOL items may have radically different components than next-generation electronics. Due to the continually evolving nature of electronic devices, the majority of waste PSCs' components (such as the ETL, perovskite layer, HTL, and back electrode) cannot be frequently recycled.

Conflict of interest

None

Acknowledgment

Authors (TKG, SJ, and KC) extend their thanks and appreciation to Amity Institute of Applied Sciences, Amity University Uttar Pradesh, Noida, India for their constant support and encouragement throughout of this work. One of the authors Kalpana Lodhi is grateful to Director, CSIR-National Physical Laboratory, New Delhi, India, and Dr Sushil Kumar (Head), Photovoltaic Metrology Section, CSIR-National Physical Laboratory, New Delhi, India, for support and valuable guidance. Manjeet Singh Goyat is grateful to Director, University of Petroleum & Energy Studies, Dehradun 248007, Uttarakhand, India.

References

[1] A. Kojima, K. Teshima, Y. Shirai, T. Miyasaka, Organometal halide perovskites as visible-light sensitizers for photovoltaic cells, J. Am. Chem. Soc. 131 (2009) 6050-6051. https://doi.org/10.1021/ja809598r

[2] M. Liu, M.B. Johnston, H.J. Snaith, Efficient planar heterojunction perovskite solar cells by vapour deposition, Nature. 501 (2013) 395-398. https://doi.org/10.1038/nature12509

[3] F. Yang, J. Liu, Z. Lu, P. Dai, T. Nakamura, S. Wang, L. Chen, A. Wakamiya, K. Matsuda, Recycled utilization of a nanoporous Au electrode for reduced fabrication cost of perovskite solar cells, Adv. Sci. 7 (2020) 1902474. https://doi.org/10.1002/advs.201902474

[4] J.I. Bilbao, G. Heath, A. Norgren, M.M. Lunardi, A. Carpenter, R. Corkish, PV Module Design for Recycling Guidelines, National Renewable Energy Lab. (NREL), Golden, CO (United States), 2021. https://doi.org/10.2172/1832877

[5] E. Masanet, A. Horvath, Assessing the benefits of design for recycling plastics in electronics: A case study of computer enclosures, Mater. Des. 28 (2007) 1801-1811. https://doi.org/10.1016/j.matdes.2006.04.022

[6] G. Roos, Business model innovation to create and capture resource value in future circular material chains, Resources. 3 (2014) 248-274. https://doi.org/10.3390/resources3010248

[7] Y. Deng, Z. Ni, A.F. Palmstrom, J. Zhao, S. Xu, C.H.V. Brackle, X. Xiao, K. Zhu, J. Huang, Reduced self-doping of perovskites induced by short annealing for efficient solar modules, Joule 4 (2020) 1949-1960. https://doi.org/10.1016/j.joule.2020.07.003

[8] Y. Jiang, L. Qiu, E.J.J.- Perez, L.K. Ono, Z. Hu, Z. Liu, Z. Wu, L. Meng, Q. Wang, Y. Qi, Reduction of Lead leakage from damaged Lead halide perovskite solar modules using self-healing polymer-based encapsulation, Nat. Energy 4 (2019) 585-593. https://doi.org/10.1038/s41560-019-0406-2

[9] C.C. Stoumpos, C.D. Malliakas, M.G. Kanatzidis, Semiconducting Tin and Lead Iodide perovskites with organic cations: Phase transitions, high mobilities, and near-infrared photoluminescent properties, Inorg. Chem. 52 (2013) 9019-9038. https://doi.org/10.1021/ic401215x

[10] B.P. Dhamaniya, P. Chhillar, A. Kumar, K. Chandratre, S. Mahato, K.P. Ganesan, S.K. Pathak, Orientation-Controlled (h0l) PbI2 crystallites using a novel Pb-precursor for facile and quick sequential MAPbI3 Perovskite deposition, ACS Omega. 5 (2020) 31180-31191. https://doi.org/10.1021/acsomega.0c04483

[11] W.R. Mateker, M.D. McGehee, Progress in understanding degradation mechanisms and improving stability in organic photovoltaics, Adv. Mater. 29 (2017) 1603940. https://doi.org/10.1002/adma.201603940

[12] A. Binek, M.L. Petrus, N. Huber, H. Bristow, Y. Hu, T. Bein, P. Docampo, Recycling perovskite solar cells to avoid lead waste, ACS Appl. Mater. Interfaces. 8 (2016) 12881-12886. https://doi.org/10.1021/acsami.6b03767

[13] A. Babayigit, A. Ethirajan, M. Muller, B. Conings, Toxicity of organometal halide perovskite solar cells, Nat. Mater. 15 (2016) 247 251. https://doi.org/10.1038/nmat4572

[14] P. Billen, E. Leccisi, S. Dastidar, S. Li, L. Lobaton, S. Spatari, A.T. Fafarman, V.M. Fthenakis, J.B. Baxter, Comparative evaluation of Lead emissions and toxicity potential in the life cycle of Lead halide perovskite photovoltaics, Energy 166 (2019) 1089-1096. https://doi.org/10.1016/j.energy.2018.10.141

[15] G. Wang, Y. Zhai, S. Zhang, L. Diomede, P. Bigini, M. Romeo, S. Cambier, S. Contal, N.H. Nguyen, P. Rosická, An across-species comparison of the sensitivity of different organisms to Pb-based perovskites used in solar cells, Sci. Total Environ. 708 (2020) 135134. https://doi.org/10.1016/j.scitotenv.2019.135134

[16] A. Alassali, D. Barouta, H. Tirion, Y. Moldt, K. Kuchta, Towards a high quality recycling of plastics from waste electrical and electronic equipment through separation

of contaminated fractions, J. Hazard. Mater. 387 (2020) 121741.
https://doi.org/10.1016/j.jhazmat.2019.121741

[17] L. Huang, Z. Hu, J. Xu, X. Sun, Y. Du, J. Ni, H. Cai, J. Li, J. Zhang, Efficient
electron-transport layer-free planar perovskite solar cells via recycling the FTO/glass
substrates from degraded devices, Sol. Energy Mater. Sol. Cells. 152 (2016) 118-124.
https://doi.org/10.1016/j.solmat.2016.03.035

[18] B. Augustine, K. Remes, G.S. Lorite, J. Varghese, T. Fabritius, Recycling perovskite
solar cells through inexpensive quality recovery and reuse of patterned Indium Tin
oxide and substrates from expired devices by single solvent treatment, Sol. Energy
Mater. Sol. Cells. 194 (2019) 74-82. https://doi.org/10.1016/j.solmat.2019.01.041

[19] B.J. Kim, D.H. Kim, S.L. Kwon, S.Y. Park, Z. Li, K. Zhu, H.S. Jung, Selective
dissolution of halide perovskites as a step towards recycling solar cells, Nat. Commun.
7 (2016) 1-9. https://doi.org/10.1038/ncomms11735

[20] J.M. Kadro, N. Pellet, F. Giordano, A. Ulianov, O. Müntener, J. Maier, M. Grätzel,
A. Hagfeldt, Proof-of-concept for facile perovskite solar cell recycling, Energy
Environ. Sci. 9 (2016) 3172-3179. https://doi.org/10.1039/C6EE02013E

[21] L. Huang, J. Xu, X. Sun, R. Xu, Y. Du, J. Ni, H. Cai, J. Li, Z. Hu, J. Zhang, New
films on old substrates: Toward green and sustainable energy production via recycling
of functional components from degraded perovskite solar cells, ACS Sustain. Chem.
Eng. 5 (2017) 3261-3269. https://doi.org/10.1021/acssuschemeng.6b03089

[22] B. Hailegnaw, S. Kirmayer, E. Edri, G. Hodes, D. Cahen, Rain on
Methylammonium Lead Iodide based perovskites: Possible environmental effects of
perovskite solar cells, J. Phys. Chem. Lett. 6 (2015) 1543-1547.
https://doi.org/10.1021/acs.jpclett.5b00504

[23] I.R. Benmessaoud, A.-L.M. Mellier, E. Horváth, B. Maco, M. Spina, H.A. Lashuel,
L. Forró, Health hazards of Methylammonium Lead Iodide based perovskites:
Cytotoxicity studies, Toxicol. Res. 5 (2016) 407-419.
https://doi.org/10.1039/C5TX00303B

[24] B. Chen, C. Fei, S. Chen, H. Gu, X. Xiao, J. Huang, Recycling Lead and transparent
conductors from perovskite solar modules, Nat. Commun. 12 (2021) 1-10.
https://doi.org/10.1038/s41467-020-20314-w

[25] F. Yang, J. Liu, X. Wang, K. Tanaka, K. Shinokita, Y. Miyauchi, A. Wakamiya, K.
Matsuda, Planar perovskite solar cells with high efficiency and fill factor obtained

using two-step growth process, ACS Appl. Mater. Interfaces. 11 (2019) 15680-15687. https://doi.org/10.1021/acsami.9b02948

[26] H. Zhang, J. Xiao, J. Shi, H. Su, Y. Luo, D. Li, H. Wu, Y.B. Cheng, Q. Meng, Self-adhesive macroporous carbon electrodes for efficient and stable perovskite solar cells, Adv. Funct. Mater. 28 (2018) 1802985. https://doi.org/10.1002/adfm.201802985

[27] Z. Ku, X. Xia, H. Shen, N.H. Tiep, H.J. Fan, A mesoporous Nickel counter electrode for printable and reusable perovskite solar cells, Nanoscale 7 (2015) 13363-13368. https://doi.org/10.1039/C5NR03610K

[28] Z. Li, Y. Zhao, X. Wang, Y. Sun, Z. Zhao, Y. Li, H. Zhou, Q. Chen, Cost analysis of perovskite tandem photovoltaics, Joule 2 (2018) 1559-1572. https://doi.org/10.1016/j.joule.2018.05.001

[29] M. Cai, Y. Wu, H. Chen, X. Yang, Y. Qiang, L. Han, Cost-performance analysis of perovskite solar modules, Adv. Sci. 4 (2017) 1600269. https://doi.org/10.1002/advs.201600269

[30] Z. Song, C.L. McElvany, A.B. Phillips, I. Celik, P.W. Krantz, S.C. Watthage, G.K. Liyanage, D. Apul, M.J. Heben, A technoeconomic analysis of perovskite solar module manufacturing with low-cost materials and techniques, Energy Environ. Sci. 10 (2017) 1297-1305. https://doi.org/10.1039/C7EE00757D

[31] N.L. Chang, A.W.Y.H. Baillie, D. Vak, M. Gao, M.A. Green, R.J. Egan, Manufacturing cost and market potential analysis of demonstrated roll to-roll perovskite photovoltaic cell processes, Sol. Energy Mater. Sol. Cells 174 (2018) 314-324. https://doi.org/10.1016/j.solmat.2017.08.038

[32] A. Louwen, W.V. Sark, R. Schropp, A. Faaij, A cost roadmap for Silicon heterojunction solar cells, Sol. Energy Mater. Sol. Cells. 147 (2016) 295 314. https://doi.org/10.1016/j.solmat.2015.12.026

Perovskite based Materials for Energy Storage Devices
Materials Research Foundations 151 (2023) 111-154

Materials Research Forum LLC
https://doi.org/10.21741/9781644902738-5

Chapter 5

Lead-Free Perovskite Solar Cells

Mridula Guin[1]* and Riya Singh[1]

[1]Department of Chemistry and Biochemistry, Sharda University, Greater Noida, India

Abstract

The breakthrough in 2012 for halide perovskite solar cells (PSC) left a deep impact on the next-generation solar cell. Lead halide perovskites have remarkable optoelectronic properties and are found to be highly efficient solar cells. However, the toxic nature of lead in these devices is of serious concern and set back the pace of their large-scale commercialization as PSC. The development of environment-friendly lead-free PSCs became the prime focus for scientists at large. In a short period, reasonable progress has been accomplished in lead-free PSCs. In this chapter present status of lead-free PSCs and prospects have been discussed. A detailed discussion of different strategies that can be adopted to enhance the photovoltaic efficiency of lead-free PSCs has been presented. We have also highlighted the materials to explore further progress in this area. Lastly, different fabrication processes of high-quality PSC film and associated challenges in improving efficiency have been provided.

Keywords

Perovskite, Solar Cell, Lead-Free, Halide Double Perovskites, Tin-Based, Germanium Based, Photovoltaic

Contents

Lead-Free Perovskite Solar Cells 111

1.   Introduction...112

2.   Categories of Lead-Free Perovskite Solar Cells (PSCs) 115

    2.1   Tin-Based PSCs...117

    2.2   Germanium-Based PSCs 121

    2.3   Antimony and bismuth-based PSCs 122

2.4   Halide double perovskites (HDPs) 125

3.   **Improvement Scopes in Lead-Free PSCs** **129**
3.1   Photovoltaic Efficiency 129
3.2   Stability 131
3.3   Defect Parameter Characterization and Defect Tolerance 133
3.4   Charge Transport Characterization 133
3.5   Electronic Dimensionality 134

4.   **Processing of High-Quality Lead-Free Perovskite Films** **136**
4.1   Vapour deposition method 136
4.2   Anti-Solvent Technique 136
4.3   Solution Processing 136
4.4   Two-Step Deposition 137
4.5   Low Pressure Assisted Solution Processing 137
4.6   Spin Coating 138
4.7   Inter-diffusion Method 139
4.8   Doctor Blade Coating 139
4.9   Vacuum Flash-Assisted Solution Process (VASP) 139
4.10   Complex Assisted Gas Quenching (CAGQ) method 139
4.11   Soft Cover Deposition (SCD) 139

Conclusion and outlook **139**

References **140**

1.   Introduction

The requirement for energy has skyrocketed because of rapid industrial and economic development with a growing population. Traditional fossil fuels are insufficient to meet the energy demands of today's generation. At this stage, solar energy is the best option for clean and renewable nature. The conversion of solar energy into electronic energy using optoelectronic or photovoltaic materials has become a subject of intense interest in recent years. In the field of photovoltaics, perovskite solar cells (PSCs) are at the forefront because of their excellent absorption coefficient, high mobility of charge carriers, the low binding energy of exciton, long diffusion length, adjustable band gap, and low trap density states. Among these, lead-halide-based perovskites (LHPs) have received tremendous attraction from the scientific community across the globe [1-6]. These inorganic-organic hybrid perovskites (IOHPs) have phenomenal optoelectronic properties. They have a 25 times higher absorption coefficient than that of Si and GaAs materials. The power conversion efficiency has reached from early value of 3.8% to more than 25% [7]. The straightforward manufacturing process of lead halide perovskites is the main reason behind their successful rise. However, there are three important issues with lead halide which prevent their commercial growth. The first challenge is performance degradation due to strong reactivity with oxygen and moisture [8-11]. The second issue is structural instability because of crystallization into unwanted photo-inactive phases [12,13]. And the last one is related to the impact of the toxicity of lead halide on the environment [14-16]. However, the instability issue has been solved recently [17], and tackling the toxic lead remains the primary challenge. Thus researchers are looking for alternatives for lead-free perovskites having similar optoelectronic properties as that of lead halide perovskites [18]. The alternative lead-free absorbers must have direct band gaps for effective light absorption and photon recycling. It should have small and balanced effective masses for electrons and holes for efficient ambipolar transport. It must be defected tolerant to suppress nonradiative recombination channels. Lastly, these compounds should have easy processing technology to compete with established PV materials [19-22].

Typically two approaches have been applied to develop non-toxic or low-toxic perovskite photovoltaics. In this regard, the complete substitution of lead with absorbers based on metal ions such as tin(II), germanium (II), antimony(II), bismuth(II), copper(II), and manganese(II) [23-25] and partial substitution of lead with non-toxic metal ions have been investigated [26-30]. The replacement of lead with other possible elements introduces diversity in lead-free perovskites for a greener and safe environment. Density Functional Theory (DFT) calculations, keeping stability and optimum band gap as a prerequisite by Filip et al. screened out suitable elements for replacing lead [31] [Fig.1]. Within a very short time, considerable progress has happened reaching up to 13.2% PCE in this area. The first strategy implemented is the substitution of lead with elements with ns^2 configuration to develop lead-free PSC. Perovskites with bivalent metallic elements have the advantage of a considerable charge transfer process. In this regard, Sn^{2+} was the first choice due to its similar ionic radius and electronic configuration [32-34]. Unfortunately, the oxidation of Sn^{2+} to Sn^{4+} restricts its application as an alternative lead-free PSC. Moreover, tin perovskites generate SnI_2, a hazardous substance similar to PbI_2. Next, Ge^{2+} as a divalent alternative to lead could not stand out due to its low PCE value. Wide band gap, smaller ionic radius, and low solubility are the reasons for the low PCE of the quasi 3D-hexagonal structure of Ge-based perovskites [35,36]. Ge^{4+} being more stable than Ge^{2+}, there is a stability issue in the perovskite lattice structure. The smaller ionic radius of copper leads to a low adsorption coefficient and low intrinsic conductivity in Cu-based perovskites. The result of the computational investigation to find out homovalent ions to replace lead confirms that no divalent metal ion is as powerful as that of lead and tin-based perovskites. Thus heterovalent substitution of lead is the only option left out. Heterovalent elements e.g. antimony, bismuth possessing a similar electronic configuration as Pb^{2+} will be a wise choice.

Because of the heterovalent substitution, the electrical charge neutrality is maintained by altering the perovskite crystal structure. Among several options for heterovalent replacement, the first one is to substitute Pb^{+2} with trivalent cations e.g. Sb^{+3}, $Bi^{+3,}$ etc. In that case, the ABX_3 perovskite lattice accompanied a structural transformation into ABX_3 to $A_3B_2X_9$. The other way is a halide double perovskite (HDP) structure with general formula $A_2B(I)B'(III)X_6$, where B(I) and B'(III) are monovalent and trivalent cations respectively. This is called the cation-splitting method. Another form of HDP can be prepared by ordered vacancy with the general formula $A_2 \cdot B(IV)X_6$, where B is a tetravalent cation along with vacancy in the B site (represented by sign •). In the case of B(III) compounds, an ordered vacancy can be created by $A_3 \cdot B(III)X_9$ lattice structure. However, the efficiencies of these perovskites did not reach up to the mark because of the low electronic dimension. The presence of vacancies diminishes the optoelectronic properties for low performance but attempts are still going on for improving their efficiency. Therefore scientists turn to a mixed-valence anion approach to prepare chalcogenide perovskites with the general formula $AB(Ch, X)_3$ which has excellent stability. In this case, there remains a single valence cation but two valence anions represented by Ch, a chalcogen, and X, a halogen element. But the preparation of chalcogenide perovskites requires very high temperatures putting constraints on their widespread application in the solar cell. Therefore, the idea is to devise a synergic combination of highly stable metal chalcogenides with highly efficient metal halide perovskite solar cells. These mixed chalcogenides and halides are expected to be promising to obtain lead-free stable highly efficient PSC.

In this chapter, we highlighted the current status and perspective of lead-free perovskite absorbers. Different types of lead-free absorbers showing promising results and challenges involved in increasing photovoltaic efficiency have been discussed. Moreover, a strategy to overcome the challenges is also outlined. It has been observed that the deficit of experimental optoelectronic data on lead-free PSCs is the reason behind the slow pace of improving their performance and development.

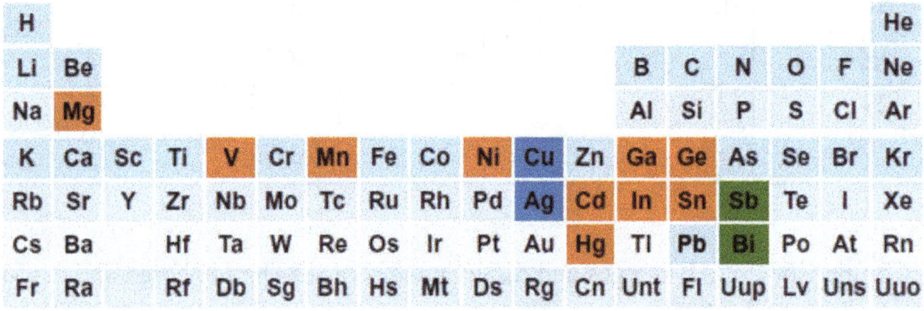

Materials Research Forum LLC
https://doi.org/10.21741/9781644902738-5

Figure 1. Potential elements (colored with orange, green and blue) to substitute Pb determined from the computational investigation.[71]

2. Categories of Lead-Free Perovskite Solar Cells (PSCs)

A lot of research has been published on lead-free perovskite absorbers. The most investigated materials have been tin, germanium, antimony, or bismuth-based and double perovskites photovoltaics. Table 1 summarizes the comparison of a few important parameters of lead and other possible replaceable metal atoms. Moreover, in Table 2 advantages and disadvantages of lead-based PSC as well as that of various types of lead-free PSCs are listed. In the next section, different types of lead-free PSCs are discussed in detail

Table 1. Comparison of some parameters of Pb and other possible replacement metals

Element	Atomic No	Electronic configuration	Oxidation states	Ionic Electronic configuration	Ionic radius (pm)
Pb	82	$[Xe]4f^{14}5d^{10}6s^26p^2$	+2 (stable)	$[Xe]4f^{14}5d^{10}6s^2$	119
			+4	$[Xe]4f^{14}5d^{10}$	77
Sn	50	$[Kr]4d^{10}5s^25p^2$	+2	$[Kr]4d^{10}5s^2$	118
			+4(stable)	$[Kr]4d^{10}$	69
Ge	32	$[Ar]3d^{10}4s^24p^2$	+2	$[Ar]3d^{10}4s^2$	73
			+4	$[Ar]3d^{10}$	53
Bi	83	$[Xe]4f^{14}5d^{10}6s^26p^3$	+3(stable)	$[Xe]4f^{14}5d^{10}6s^2$	103
			+5	$[Xe]4f^{14}5d^{10}$	90

Sb	51	$[Kr]4d^{10}5s^25p^3$	+3	$[Kr]4d^{10}5s^2$	76
			+5	$[Kr]4d^{10}$	62
Cu	29	$[Ar]3d^{10}4s^1$	+1	$[Ar]3d^{10}$	77
			+2	$[Ar]3d^9$	73

Table 2. Comparisons of advantages and disadvantages of lead halide based and lead-free PSC

Type of Perovskite solar cell (PSC)	Advantages	Disadvantages
Lead halide based PSC ($APbX_3$)	1. High optical absorption coefficient 2. Small carrier effective masses 3. Excellent defect tolerance 4. High electronic dimension	1. Instability in long term 2. Lead toxicity
Lead free PSC (isovalent replacement)		
$AB(II)X_3$	1. Electronically 3 dimension 2. High-performance	Stability issue
$A_2B(II)X_4$	Good stability	1. High defects 2. Low absorption coefficient
$A \bullet B(III)_2X_9$		
Lead free PSC (heterovalent replacement)	**Advantages**	**Disadvantages**
$AB(Ch, X)_3$	Good stability	1. High defects 2. Wide band gap
$A_2 \bullet B(IV)X_6$	Electronically 3 dimension	Thermodynamically unstable
$A_2B(I)B(III) X_6$	1 Good stability 2. Appropriate band gap	1. High defects 2. Large hole effective mass

2.1   Tin-Based PSCs

The general formula of lead perovskite material is ABX_3 in which A is a cation, e.g. Cs^+, CH_3NH_3, NH_2CHNH2; B^{2+} is another cation Pb, and X is a halide ion which can be chloride, bromide, iodide or a mixture of halides. The photovoltaic efficiency of lead perovskites is comparable to the well-known polycrystalline silicon solar cell. This is due to the remarkable photophysical and morphological properties of these compounds. Especially in the case of iodide, the orbital overlapping between lead and iodine is very optimal. Even though lead halide perovskites are the top performers in terms of performance and device qualities, tin halide-based perovskites are recently gaining a lot of interest. When the lead was substituted by tin partially or completely, numerous beneficial features have been obtained. As lead-based perovskites had already set the mark high enough with their excellent photovoltaic properties and exceptionally high PCE, the first one that is considered for replacing toxic lead is tin. There is a list of reasons behind considering tin as a better replacement for lead [37,38]. Tin has superior optoelectronic properties compared to lead. Both of them belong to the same 4th (IV A) group of the periodic table. Moreover, tin is considered to be non-toxic. Tin-based perovskite showed almost similar optical band gap as lead, along with that, the band gap is also close to the Shockley-Queisser limit. The preparation techniques of tin halide perovskites are easy, adaptable, and possess other advantages such as low exciton binding energy and high charge carrier mobility. Formation of 3-dimensional ABX_3 (where A is a monovalent cation, B is Sn and X halide) perovskites which structurally closely matches with commercial LHPs (Fig.2a), is possible because of the comparable ionic radius of Sn^{2+} and Pb^{2+}. Tin-based perovskites exhibit a band gap of 1.4 eV, large carrier mobility, high absorption coefficient, and large production of photocurrent density. That's the reason why tin-based perovskites are an important alternative for eco–friendly and lead-free perovskite and are now a hot topic of research for the scientific community.

Despite these advantages of tin-based perovskites, the problem lies in its low open circuit voltage (Voc) which is much lesser when compared to lead-based perovskite. The reason behind the low open circuit voltage is high p-type doping resulting in the oxidation of Sn^{2+} to Sn^{4+}, which in turn results in the recombination of photocarriers and ultimately decreases the PCE [39-42]. Thus, the disadvantage of tin-based perovskite is its instability due to oxidation. Additionally, high morphological defect density lowers the Voc value of these materials. Another issue is the degradation product, SnI_2 is as hazardous as PbI_2 putting a question mark on environmental safety. The commonly studied tin-based perovskites are formamidinium tin iodide ($FASnI_3$), cesium tin iodide ($CsSnI_3$), and methylammonium tin iodide ($MASnI_3$), having bandgaps approximately as 1.41, 1.3 and 1.2 eV respectively. These band gaps are actually narrower than lead-based perovskite and are ideal for single-junction photovoltaics. In many cases, these materials show a high short circuit current (more than 20 mA cm^{-2}). The main concern of these materials is their instability and low PCE value (less than 7%). Thus, the main focus of using tin-based perovskite solar material is to improve its stability and photophysical properties. As Sn(II) consists of ionic crystals as halide perovskite, it gets difficult to maintain stability due to penetrating water and oxygen. Even though its issues with stability, tin with a similar ionic radius as lead found its advantage and a chance to get explored first for lead-free perovskite. The instability issue is mostly resolved by using reducing agents (e.g. SnF$_2$) as additives leading to an improved PCE of about 10% [43-46]. The enhancement of its efficiency is now on focus by applying different alternative approaches. One such approach is to manipulate the perovskite structure of tin-based perovskites for better stability. To stabilize the tin-based perovskites, attempts are made to develop derivatives of tin halide perovskite of the general formula A$_2$SnX$_6$. These compounds have great stability due to the presence of the Sn^{4+} state. The lattice structure can be imagined as if derived from Cs$_2$Sn$_2$X$_6$ by eliminating one-half of Sn atoms from the octahedral position. These are termed vacancy-ordered double perovskite structures (For detail see Sec. 2.4). Although these perovskites have higher stability but suffer from low PCE value. Another method is to develop hollow perovskites by producing structural voids using medium-sized cations (Fig.2b)[47,48]. This resulted in improved stability and enhanced PCE value. Further, manipulation of dimensionality can also improve PCE value. When the 3D domains of perovskites are blended with quasi-2D domains in a layered form, the PCE value improved significantly [49-51]. The organic-inorganic perovskite MASnI$_{3-x}$Br$_x$ showed a tunable bandgap between 1.3 and 1.75 eV by varying the composition of iodide as x= 0 to 2. As bromine is added, the V$_{oc}$ improved to a good extent. A yield of up to 6.4% PCE has been achieved in this case. Other than the oxidation problem in the atmosphere from Sn^{2+} to Sn^{4+}, optoelectronic properties are highly promising for tin-based perovskites such as narrow bandgap, high electron mobility, and long charge carrier diffusion length. To solve this problem SnF$_2$ as an additive is incorporated and this has been used widely by tin perovskite researchers to attain the stability of Sn^{2+}. Another type of tin perovskite that has been investigated is FA-based Sn-perovskite(FASnI$_3$) which showed better results than MASnI$_3$. It is due to its improved surface morphology, good electrical properties, and wider band gap compared to MASnI$_3$. One interesting fact is that FASnI$_3$ is stable at room temperature, whereas FAPbI$_3$ is thermodynamically unstable. FASnI$_3$ became the popular choice for tin-based perovskite rather than MASnI$_3$ because of its better atmospheric stability. Table 3 lists electron mobilities and hole mobilities of recently reported tin halide-based absorbers. Fig.3 shows the progress in the efficiency of tin-based perovskite solar cells [52].

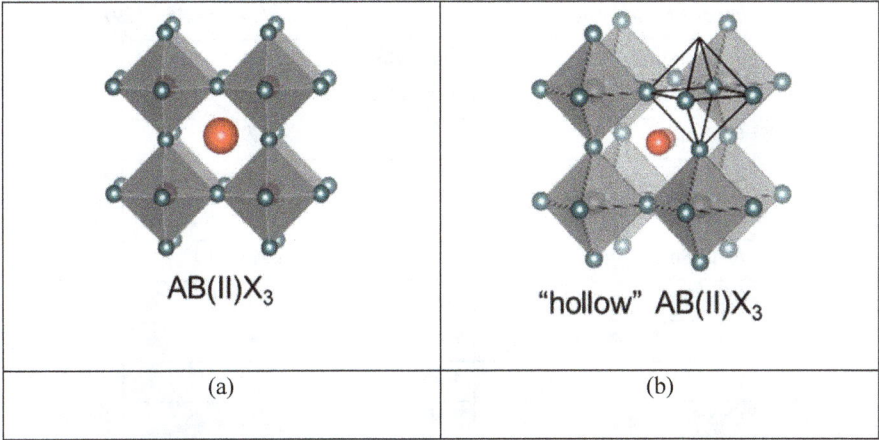

Figure 2. Common crystal structure of lead-free perovskites [55]

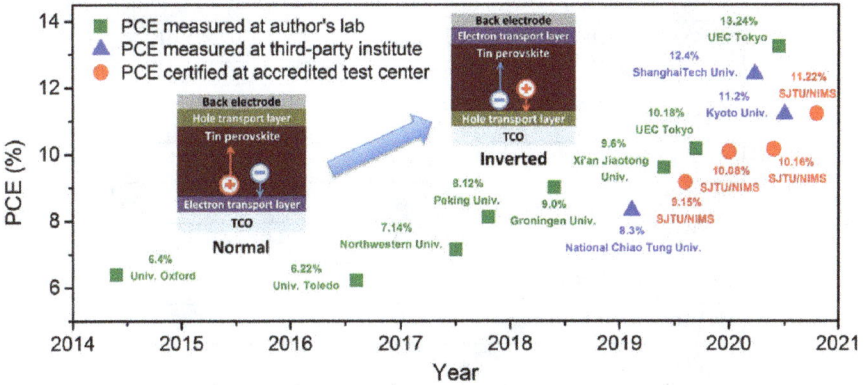

Figure 3. Power conversion efficiency (PCE) of tin-based PSCs from year 2014 to 2020 [52]

Table 3. Photophysical properties of tin halide-based perovskites [19]

Perovskite absorber	Electron mobility $(cm^2\ V^{-1}s^{-1})$	Hole mobility $(cm^2\ V^{-1}s^{-1})$
Cs_2SnI_6	2.9	2.9
#$FASnI_3$	1.6	1.6
$MASnI_3$	1.6	1.6
#$MASnI_3$	1.6×10^{-9}	1.6×10^{-9}
#$CsSnI_3$	2267.84	330.80
$FA_{0.75}MA_{0.25}Sn_{1-x}Ge_xI_3$	98.27	98.27
$(CH_3NH_3)_2SnI_6$	3.05 ± 0.56	3.05 ± 0.56
$CsSn_{0.5}Ge_{0.5}I_3$	974	213
$CsSnIBr_2$	2.5×10^{-4}	2.5×10^{-4}
$CH_3NH_3SnI_3$	1.6	1.6
$(PEA)_2(FA)_8Sn_9I_{28}$	1.055×10^5	1.055×10^5
Cs_2SnI_6	310	42
$FASnI_3$	7.2×10^{-1}	2.5×10^{-2}
$BA_2MA_3Sn_4I_{13}$	7.2×10^{-3}	6.8×10^{-3}
#$MASnI_3$	1.6	1.6
Ag_8SnS_6	14- 23	14-23
$MA(Pb_{0.3}Sn_{0.7})Br_3$	20	20
$(FASnI_3)_{0.6}(MAPbI_3)_{0.4}$	-	40 ± 5
Cs_2SnI_6	310	310
$Cs_{0.375}FA_{0.625}SnI_3$	2.66×10^{-4}	2.66×10^{-4}
$MAPb_{0.9}Sn_{0.05}Cu_{0.05}I_{2.9}Br_{0.1}$	0.83	1.20
$CH_3NH_3SnI_3$	2000	300
#$CH_3NH_3SnI_3$	5445	14 470
$FASnI_3$	22	22

2.2   Germanium-Based PSCs

Germanium is one of the attractive candidates from group 14 to take the seat in the lead-free perovskite solar cell. Germanium is being explored to replace lead, because of its divalent nature as germanium possesses similar optical, transport properties and electronic similarities. This allows the formation of $AGeX_3$ (where A is a monovalent cation and X a halide) perovskite structure. The outer electronic structure of $Ge^{2+}(4s^2)$ is similar to that of $Sn^{2+}(5s^2)$ and $Pb^{2+}(6s^2)$, but the ionic radius is smaller as compared to the other two. The most interesting candidate belonging to the family of $AGeX_3$ is $CsGeI_3$ due to its direct band gap [53]. The use of other A-site cations or halogens leads to a change in the band gap. For example, the use of lighter halides makes the band gap wider [53,54]. Also if another bulky group cation is used then not only does the bandgap increase but also the nature is changed as well from direct to indirect. In the $CsGeX_3$ family, $CsGeI_3$ has the narrowest bandgap of about 1.63 eV. As we keep on replacing the X with other halides it is seen that the bandgap keeps on increasing with the decrease in size of the halide ion. In the case of $CsGeX_3$, if X is chosen as I, Br, or Cl then the band gap obtained is 1.6, 2.3, and 3.2 eV respectively. The band gap indicates that among these only $CsGeI_3$ has the suitable bandgap to be used as a photovoltaic material and as a good absorber. In theory, the $CsGeI_3$ efficiency (~30%) is in the Shockley-Queisser limit [55] [Fig.4], however, the research on germanium-based perovskite solar cells is very less to date. It has been observed that Ge- based perovskite has less given attention than Sn-based perovskite. Thus, making the Ge-based perovskite to be stable and efficient with high performance turned out to be even more challenging. This is due to the instability of germanium to get oxidized from Ge^{2+} to Ge^{4+} state and also due to the low binding energy $4s^2$ electron of germanium. Due to low stability and low PCE, the use of germanium perovskite in the photovoltaic application is limited.

To overcome these challenges in Ge-based perovskites, the use of two metals together e.g. tin and germanium are one of the best strategies. In this new promising alternative Sn-Ge alloy ($ASn_{1-x}Ge_xX_3$) proved to be stable and also has a narrow bandgap. In the theoretical study on the possible Sn-Ge perovskite, many combinations came out to be promising. One such perovskite is $RbSn_{0.5}Ge_{0.5}I_3$ with a direct band gap of 1.6 eV. Also, the optical absorption of this is similar to that of lead perovskite. By varying the ratio of Sn-Ge, the optoelectronic properties can be tuned. Some progress made in this direction includes $FA_{0.75}MA_{0.25}Sn_{0.95}Ge_{0.05}I_3$, with a band gap of 1.4 eV [56]. The efficiency of the device was improved when the cells were kept in a nitrogen atmosphere for up to 72 hours. This stability was due to the low disorder effect arising from alloying and passivation combinations of germanium Ge^{2+}. Not only this, but the device was able to retain 80% of the efficiency in the outer atmosphere and support its stability well. This recent result was shown by bimetallic perovskite of general formula $FA_{0.75}MA_{0.25}Sn_{1-x}Ge_xI_3$ exhibiting enhanced stability with good efficiency with PCE of 7.9% due to trap healing effect. Another promising result was seen in the case of $CsSn_{0.5}Ge_{0.5}I_3$, with a PCE of 7.11%, with an improvement in stability [57]. The bandgap obtained was 1.5 eV. The solar cell can maintain up to 92% of its efficiency even after the continuous work of 500 h in a nitrogen atmosphere. Whereas when exposed to air, it was still able to maintain up to 91% of its efficiency even after continuous service of 100 hours. This improvement was obtained due to one thin layer of protection of GeO_2 on $CsSn_{0.5}Ge_{0.5}I_3$ film.

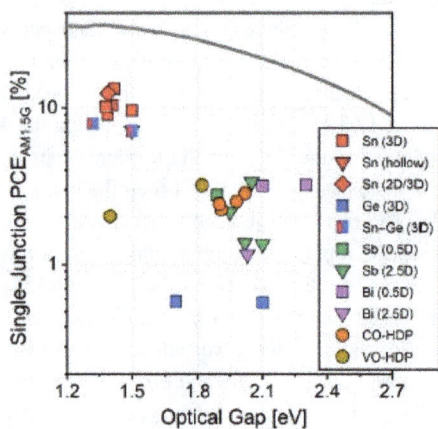

Figure 4. Power conversion efficiency (PCE) of single junction solar cell under AM 1.5G illumination against band gap of different perovskite absorbers(solid symbols). Also, theoretical PCE under AM 1.5G illumination in the Shockley-Queisser limit is shown (solid line).[55]

2.3 Antimony and bismuth-based PSCs

In the race of finding the perfect material to replace lead to make lead-free perovskite solar cells, antimony and bismuth as options came out as an attractive solutions for this replacement, mainly due to their low toxicity [58]. These are being explored as higher and lower dimensional perovskite materials, with good stability. Other than that, the electronic configuration and the ionic radii of Sb and Bi are similar to Pb. The VA group element antimony and bismuth have the same electronic configuration as the 3+ cations with Pb^{2+}. Thus the PV performance for these newly explored materials is expected to be similar to lead-based perovskite solar cells.

Perovskite based Materials for Energy Storage Devices
Materials Research Foundations 151 (2023) 111-154

Materials Research Forum LLC
https://doi.org/10.21741/9781644902738-5

For antimony-based perovskite material, chalcogenides came out as a possible composition to be used in solar cells. The highest PCE reported for these materials is 6.6%. The composition of the absorber as $A_3B_2X_9$, where B^{3+} is Sb^{3+} or Bi^{3+}, can be present in 2 phases [59-60]. First, as a dimer phase (0.5D) consisting of face-sharing metal halide bi-octahedra [Fig.5a]. The second one is a layered phase (2.5D) with planes of corner-sharing metal halide octahedra [Fig.5b].

This composition of perovskite $Bi_3I_9^{3-}$ consists of 2 face-sharing octahedral, where the A^+ cation occupies the space in-between and makes it a zero-dimensional perovskite. In one of the studies, it is seen that $MA_3Bi_3I_9$ is comparatively more stable than $MAPbI_3$. Also, it is noted that MBI is not degraded to form BiI_3 in presence of moisture in the air. It does so by forming a surface layer that does not let the recombination rate increase. The first report based on Bi-based perovskite solar cells has given 0.12% efficiency of MBI and 1.09% efficiency of $Cs_3Bi_3I_9$ [61].

Researchers are focusing on 0.5D ternary iodide $A_3B_2I_9$ (where A is Cs^+ or MA^+) as an absorber in recent investigations [62-64]. The band gap of this absorber falls in the region of 2.1-2.4 eV indicating its potential in the field of tandem photovoltaics. However, recent research is mainly focused on single-junction PV cells for 0.5D antimony and bismuth-based absorbers. The developments in this area consist of a building on additive uses optimized transport layers or dedicated deposition protocols. These developments have significantly improved the PCE reaching up to 3.2% [65,66]. Also in some cases, it showed excellent device stability in outer air and was able to retain 97% of the initial PCE value for over two months. The higher dimension of antimony and bismuth-based absorbers e.g. 2.5D perovskites $A_3B_2X_9$ makes them more suited for the job. The advantage is due to the small exciton binding energy, narrower band gaps, and effective masses. An efficiency of 1.4% has been reported for 2.5D $Rb_3Sb_2I_9$ with both the planar and mesoporous device structures [67, 68]. Recently, efficiencies of 3.34 and 2.2% have been achieved with 2.5D $MA_3Sb_2Cl_xI_{9-x}$ and $Cs_3Sb_2Cl_xI_{9-x}$ respectively [69,70]. Performance parameters of some recently reported important bismuth-based PSCs are described in Table 4. [71]

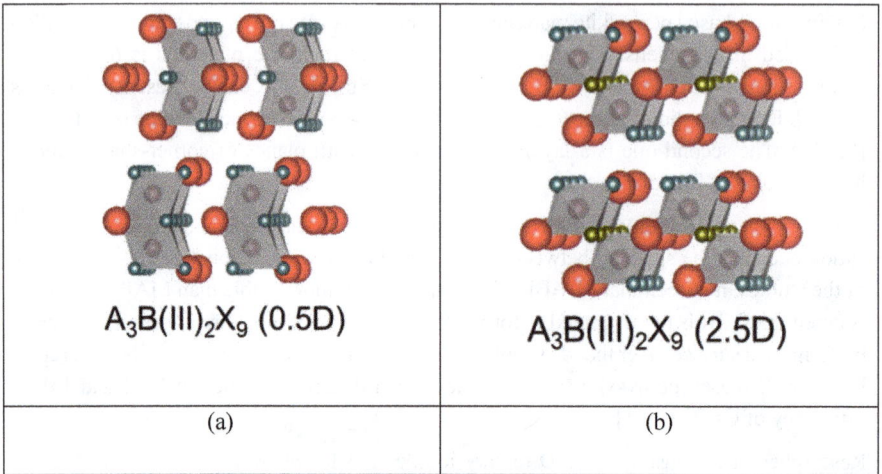

$A_3B(III)_2X_9$ (0.5D)

$A_3B(III)_2X_9$ (2.5D)

| (a) | (b) |

Figure 5. Crystal structure of $A_3B_2X_9$ perovskites (a) dimer phase [0.5D] (b) layered phase [2.5D] [55]

Table 4. Performance parameters of bismuth-based PSCs [71]

Absorber	Preparation Process	Device Structure	V_{OC} (V)	J_{SC} (mA cm^{-2})	FF (%)	PCE (%)
$Cs_3Bi_2I_9$	One- step spin coating	c-TiO$_2$/m-TiO$_2$/ perovskite/ spiro-MeOTAD/Ag	0.85	2.15	60	1.09
$Cs_3Bi_2I_9$	One- step spin coating	c-TiO$_2$/m-TiO$_2$/ perovskite/ P3HT/Ag	0.31	0.34	38	0.40
$Cs_3Bi_2I_9$	One- step spin coating	c-TiO$_2$/ perovskite/ spiro-MeOTAD/Au	0.79	4.45	50.34	1.77
$Cs_3Bi_2I_9$	One- step spin coating	c-TiO$_2$/ perovskite/ PTAA/Au	0.83	4.82	57.49	2.30
$Cs_3Bi_2I_9$	One- step spin coating	c-TiO$_2$/ perovskite/ CuI/Au	0.86	5.78	64.38	3.20
$(CH_3NH_3)_3Bi_2I_9$	One- step spin coating	c-TiO$_2$/ perovskite/ spiro/Au	0.72	0.49	31.8	0.11

$(CH_3NH_3)_3Bi_2I_9$	One- step spin coating	c-TiO$_2$/m-TiO$_2$/ perovskite/ spiro-MeOTAD/Ag	0.68	0.52	33	0.12
$(CH_3NH_3)_3Bi_2I_9$	One- step spin coating	c-TiO$_2$/m-TiO$_2$/ perovskite/ spiro-MeOTAD/Au	0.68	0.38	88	0.22
$(CH_3NH_3)_3Bi_2I_9$	One- step spin coating	c-TiO$_2$/m-TiO$_2$/ perovskite/ PIF8-TAA/Au	0.85	1.22	73	0.71
$(CH_3NH_3)_3Bi_2I_9$	One- step spin coating	c-TiO$_2$/m-TiO$_2$/ perovskite/ P3HT/Au	0.35	1.157	46.4	0.19
$(CH_3NH_3)_3Bi_2I_9Cl_x$	One- step spin coating	c-TiO$_2$/m-TiO$_2$/ perovskite/ spiro-MeOTAD/Ag	0.04	0.18	38	0.03
$(CH_3NH_3)_3Bi_2I_9$	Two-step thermal evaporation	c-TiO$_2$/m-TiO$_2$/ perovskite/ spiro-MeOTAD/Au	0.83	3.00	79	1.64
$(CH_3NH_3)_3Bi_2I_9$	Two-step evaporation spin coating	PEDOT:PSS/perovskite/ C$_{60}$/BCP/Ag	0.83	3.00	79	1.64
$(CH_3NH_3)_3Bi_2I_9$	Vapor-assisted solution process	c-TiO$_2$/m-TiO$_2$/ perovskite/ P3HT/Au	1.01	4.02	78	3.17

2.4   Halide double perovskites (HDPs)

HDPs received considerable attention because of their appealing three-dimensional structure. HDPs are of two types: cation-ordered halide double perovskites (CO-HDPs) and vacancy-ordered halide double perovskites (VO-HDPs) [see Fig. 6(a) and 6(b)]. CO-HDPs have the general formula $A_2BB'X_6$, in which B is a monovalent cation and B' is a trivalent cation, which is positioned in the alternating places in the perovskite lattice at octahedral centers, X is a halide anion and A is monovalent cation [72]. While VO-HDPs have the general formula of A_2BX_6. The lattice structure consists of alternation in the cation B^{4+} and a vacancy of the B atom at the octahedral centers. The neighboring $[BX_6]^{2-}$ octahedral units are separated from each other [73,74]. Thus the VO-HDPs can be structurally considered quasi-zero dimensional and are different from the CO-HDPs.

Elements that can be used for making double perovskite $A_2B(\text{I})$ $B(\text{III})X_6$ crystal structure are shown in Fig. 6c.

To date, cation-ordered HDPs include many materials of different classes. Around more than 350 compounds have been synthesized and others are under process [75]. Among HDPs, research is mainly focused on silver-bismuth HDPs like Cs_2AgBiX_6 (where X is Cl or Br). But these absorbers are not the perfect pick in single-junction PV cells due to large, indirect band gaps and also the presence of the excitonic characteristic [73,76]. For example, in single junction photovoltaics, the maximum reachable efficiency taking into consideration all the above-mentioned points for $Cs_2AgBiBr_6$ is approximately 8% [77]. Even though CO-HDPs have a 3D structure, they are considered electronically zero-dimensional because of the spatial isolation of their band edges. The low-toxic nature, 3D structure, long carrier lifetimes (approximately 600 ns), and high stability of these compounds promote their interest in the research field for use as photovoltaics [78]. Moreover, their excellent stability in the air also contributed to catching the attention of photovoltaic research. The highest efficiency noted for a single-junction device is found to be somewhat between 2.2 to 2.8%. Combination with an interlayer of organic material can increase the absorption of photon on the perovskite absorber and increases the PCE value. The limitation of Cs_2AgBiX_6 pushes the researchers to work on synthesizing an alternative of this kind with a narrow band gap CO-HDP photovoltaics. The synthesis of Cs_2AgBiI_6 where iodine is used as the halide ion was challenging because of its thermodynamic instability [79]. But recently it has been synthesized as colloidal nanocrystals with a suitable band gap. This opens a door for promising alternatives for perovskite solar cell research [80]. Another interesting approach utilizes antimony in place of bismuth. In this context, important systems are $Cs_2AgSbBr_6$, (indirect band gap of 1.64 eV, PCE 0.01%) and Cs_2AgInX_6 (direct band gap along with parity forbidden transition).

Vacancy-ordered HDP has been concentrated on two kinds of perovskites mainly Cs_2SnX_6 and Cs_2TiBr_6. In Cs_2SnX_6, tin is present in oxidation state 4^+ which prevents it from degradation. In contrast to CO-HDPs, Cs_2SnX_6 has a direct band gap of 1.3-1.6 eV which enables them as an inherent absorber. The Shockley-Queisser efficiency limit for Cs_2SnX_6 is greater than 25% for single junction photovoltaic [81,82]. The highest efficiency reported is 2% with sufficient air stability. The original PCE gets reduced by only 5 % in 50 days in an encapsulated device. The other interesting VO-HDP is titanium-based Cs_2TiBr_6 with 3.3% PCE. This is an environmentally stable single junction device with a quasi-direct gap of 1.8 eV. Further research is required for A_2TiX_6 VO-HDPs to progress from their budding stage. Photovoltaic performance parameters of selected double perovskite $A_2B(\text{I})B(\text{III})X_6$ absorbers are listed in Table 5.

Materials Research Forum LLC
https://doi.org/10.21741/9781644902738-5

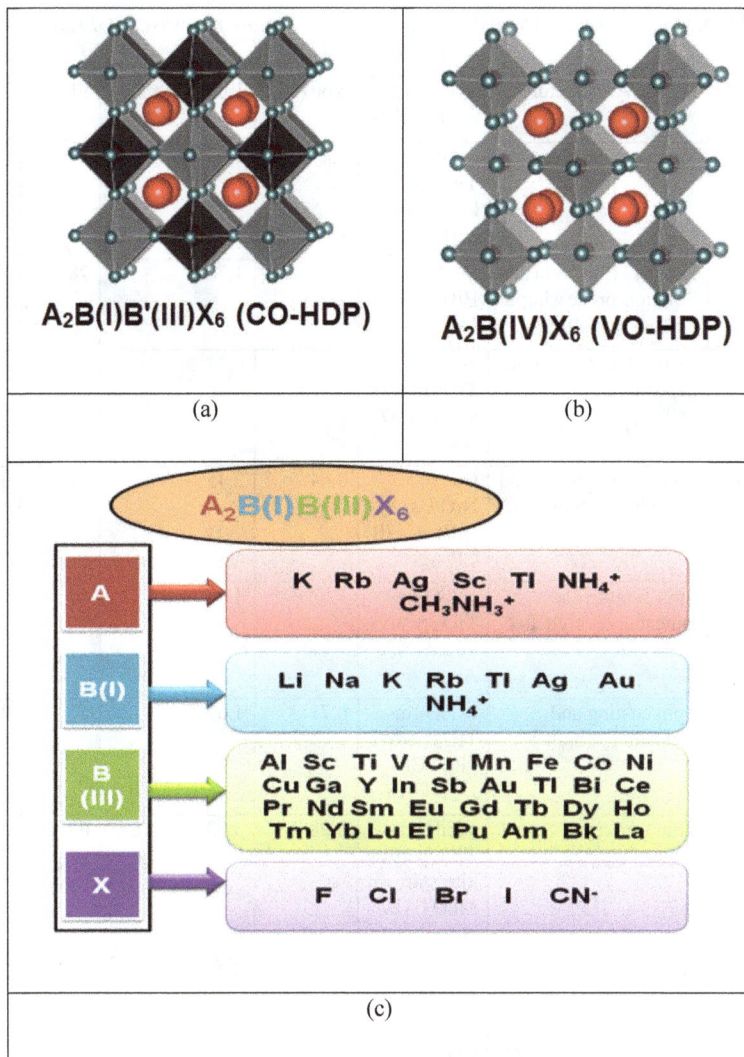

Figure 6. Crystal structure of (a) CO-HDP [55] (b)VO-HDP [55] and (c) the elements/functional groups that can form double perovskite $A_2B(I)B(III)X_6$ [71].

Table 5. Photovoltaic performance parameters of double perovskite $A_2B(I)B(III)X_6$ absorber [71]

Perovskite absorber	Device Architecture	Synthetic method #	Voc (V)	Jsc (mA cm^{-2})	FF(%)	PCE(%)
Cs_2NaBiI_6	Spin coating	c-TiO$_2$/m-TiO$_2$/perovskite/spiroMeOTAD/Au	0.47	1.99	44	0.42
$Cs_2AgBiBr_6$	Low P assisted Solution processing at ambient condition	SnO$_2$/Cs$_2$AgBiBr$_6$/P3HT/Au	1.04	1.78	78	1.44
$Cs_2AgBiBr_6$	Sequential vapour deposition	c-TiO$_2$/perovskite/P3HT/Au	-	-	-	1.37
$Cs_2AgBiBr_6$	Spin coating and annealing at high T	Cu-NiO/Cs$_2$AgBiBr$_6$/C$_{60}$/BCP/Ag	0.64	2.45	57	0.90
$Cs_2AgBiBr_6$	Spin coating and annealing at high T	c-TiO$_2$/m-TiO$_2$/perovskite/PTAA/Au	1.02	1.84	67	1.26
$Cs_2AgBiBr_6$	Spin coating and annealing at high T	c-TiO$_2$/m-TiO$_2$/perovskite/PCP-DTBT/Au	0.71	1.67	57	0.68
$Cs_2AgBiBr_6$	Spin coating	c-TiO$_2$/m-TiO$_2$/perovskite/spiroMeOTAD/Au	0.98	3.93	63	2.43
$Cs_2AgBiBr_6$	Antisolvent dropping and post anneling	Cu-NiO/Cs$_2$AgBiBr$_6$/C$_{60}$/BCP/Ag	1.00	3.23	68.4	2.21

c-TiO$_2$ and m-TiO$_2$ represent compact and mesoporous TiO$_2$ layer respectively

3.   Improvement Scopes in Lead-Free PSCs

Although lead-free perovskites solar cell has achieved considerable progress in recent years, still, improvement in stability and photovoltaic efficiency is required for their large-scale commercial applications. In this section, the challenges in the further progress of lead-free perovskites solar cell and their possible solutions have been discussed. Various factors which are closely associated with the advancement of lead-free perovskites such as charge transport, defect tolerance, dimensionality, etc. are taken into consideration for discussion in this section.

3.1   Photovoltaic Efficiency

Presently tin-based lead-free perovskites have reached the highest efficiency of 13.34% in single junction solar cell. However, their Shockley-Queisser efficiency limit (Fig. 4) is around 32% suggesting scope for further improvement. It is important to note that the short circuit current of tin-based perovskites solar cells is near the Shockley-Queisser limit. Thus it is evident that the generation of photocarriers and their collection happens efficiently. The low performance of $ASnX_3$ solar cells is due to low Voc value [20]. The reason behind low Voc is high defect density which in turn favors nonradiative recombination. This indicates that the main challenge in $ASnX_3$ solar cells is to passivate the defect density. In this regard standalone strategies won't be sufficient, a holistic approach is required for defect passivation protocols. The composition and morphology of the photoactive layer can be optimized using a variety of methods. The incorporation of additives and mixing of cations in A site are some of the ways to control the photoactive layer. In addition to that, enhancing stability by hindering the oxidation process and minimizing surface recombination also needs supervision. Efforts are needed to combat Voc deficit linked to the energetic disorder in tin-based perovskites along with the effect of composition and fabrication processing [50, 83].

Germanium-based perovskites have the lowest PCE due to low Voc and Jsc [54,84]. They undergo rapid oxidative degradation leading to instability. Therefore, $AGeX_3$ solar cells require stability enhancement protocols for future developments. On the other hand, mixed metal perovskites e.g. tin-germanium-based perovskites showed higher efficiencies (~7%) than $AGeX_3$. Improvement scopes are still lying as their PCE is much lower than the Shockley-Queisser limit. The efficiency of tin-germanium solar cells can be improved by combining the effort of stabilization against oxidation and trap passivation similar to the $ASnX_3$ system [56,57]. The Voc deficit can be boosted by controlling the energetic disorder so that improvement in overall PCE is possible.

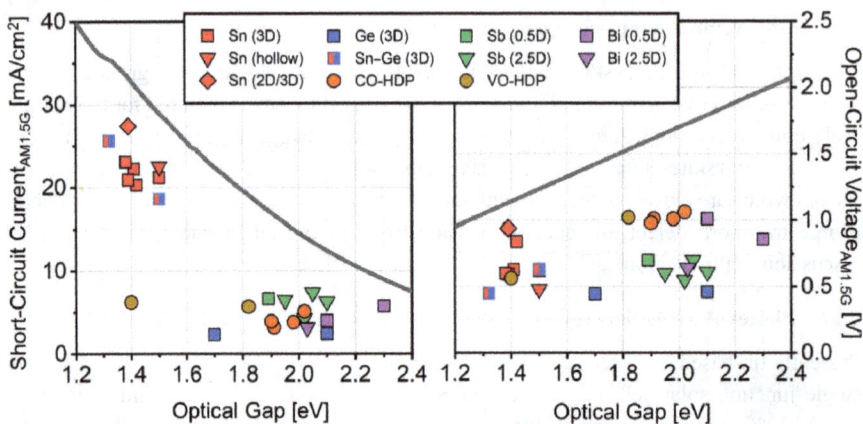

Figure 7. Plot of short circuit current (Jsc) and open circuit voltage (Voc) of reported lead free PSC under AM 1.5G illumination Also shown Jsc and Voc under AM 1.5G illumination in the Shockley-Queisser limit (solid line) for comparison [55].

In the case of both antimony and bismuth-based $A_3B_2X_9$ perovskites solar cells, the photovoltaic efficiencies are quite similar and much lower than their Shockley-Queisser limit. In the case the of 2.5D $Rb_3Sb_2I_9$ system, the Jsc value achieved was near 50% of the Shockley-Queisser limit [85]. Similarly, the reported Voc values differ significantly from the Shockley-Queisser limit [Fig. 7]. It is observed that bismuth-based systems with 0.5 dimensions are superior to antimony-based systems of both 0.5D or 2.5D categories. It is observed that the performances of 2.5D systems are better because of their efficient charge transport and defect tolerance behavior [63,86]. Antimony and bismuth-based structures in 2.5D have been investigated along the sheets of octahedra in parallel or random orientation against the substrate to understand recombination losses [69,87]. Therefore progress in deposition schemes in 2.5D $A_3B_2X_9$ systems can improve the efficiency significantly in these systems. Characterization of energetic disorder in antimony and bismuth-based perovskites are not very common. These materials are associated with high Urbach energy leading to a high Vo deficit. The energetic disorder can be tackled by refining the composition, and adjusting the deposition processes and structures of the device.

For single junction solar cell, zero-dimensional Cs_2AgBiX_6 which belongs to the CO-HDP category, have much lower efficiency than the Shockley-Queisser limit [Fig. 7]. The lower Jsc and Voc values are due to the indirect gap and the presence of excitonic effects close to the edges. Thus insufficient light absorption and low charge collection result in a decrease in photogeneration efficiency. Therefore, alternative CO-HDPs of higher dimensions and suitable optoelectronic properties need to be explored [88]. Additionally, the improvement of the photovoltaic efficiency of VO-HDPs is also to be accomplished by optimizing their synthesis procedures to control defect tolerance and charge transport property [89].

The photovoltaic efficiency of the lead-free perovskite absorbers relies on the morphology of the photoactive layer. Mostly the photoactive layers in these materials are polycrystalline and are solution deposited. Optimization of deposition conditions and methods are key factors for getting a uniform film with fewer grain boundaries. The film quality for tin-based perovskites has reached a level where they can exhibit the highest photovoltaic performance among the lead-free absorbers. However, in the case of Cs_2AgBiX_6 absorbers, the film with 400 nm grain size delivers significant efficiency [90]. $A_3B_2X_9$ absorbers based on bismuth and antimony have morphology with less than 100 nm grain size. Li et al. have demonstrated that grain size has a notable influence on the photovoltaic efficiency of $Rb_3Sb_2I_9$ absorbers [91]. Sincere efforts to establish a balance between the rates of crystallization and nucleation are necessary for the synthesis protocols to obtain desired grain size.

Finally, it can be said that the improvement of the photovoltaic performance of lead-free perovskites is possible by exploring the interfacial recombination process and optimal extraction of charge transport layers. The charge transport layers of lead-based perovskites may not be optimal for lead-free perovskites. These challenges have to be sorted out to improve the performance of these materials.

3.2   Stability

The first important criterion of a suitable lead-free halide perovskite is its inherent stability. A stable absorber is a prime force in improving the performance of perovskite solar cells. Several efforts to improve the stability of tin-based perovskites, double perovskites, antimony-based and bismuth-based perovskites have been applied. In addition to photovoltaic efficiency, commercialization of lead-free halide perovskite solar cells requires diverse methods for stability characterization. Recently standardized protocols have been proposed for the characterization of photovoltaic stability [92]. Therefore, it is worthwhile to assess the stability of the newly developed lead-free perovskites along these protocols. In this way, the stability issues can be tackled systematically to step forward for the commercialization of promising lead-free perovskite absorbers. Table 6 summarizes the efficiency and stability of tin-based perovskite absorbers with different added additives [93].

Materials Research Forum LLC
https://doi.org/10.21741/9781644902738-5

Table 6. Stability and efficiency of tin-based perovskite absorbers with different additives [93].

Perovskite Absorbers	Additive	Stability	PCE (%)
$CsSnI_3$	SnF_2	shelf life, N_2 atmosphere, 250 h (100%)	2.02
$CsSnI_3$	$SnCl_2$	25% RH, unencapsulated, 16 h (70%)	3.56
$CsSnI_3$	$SnBr_2$	40% RH, shelf stability, 100 h (98%)	4.30
$CsSnI_3$	SnI_2	N_2 atmosphere, shelf stability, 10 days (90%) .	2.76
$CsSnI_3$	$SnCl_2$, piperazine	40% RH, unencapsulated, continuous 1 sun irradiation, 30 min (40%)	2.22
$CsSnI_3$	SnF_2, CoCp2		3.0
$CsSnBr_3$	SnF_2, hydrazine	40% RH, unencapsulated, shelf stability, 5 h (40%)	3.04
$FASnI_3$	SnF_2, Sn Powder	N_2 atmosphere, unencapsulated, shelf stability, 860 h (90%)	6.75
$FASnI_3$	SnF_2, PHCl	N_2 atmosphere, unencapsulated, shelf stability, 110 d (100%)	11.4
$FASnI_3$	$SnCl_2$, AHP	20% RH, shelf stability, unencapsulated, 500 h (50%)	7.34
$FASnI_3$	SnF_2, 5-AVAI	50% RH, encapsulated, MPPT, 100 h (100%)	7.0
$FASnI_3$	SnF_2, N_2H_5Cl	N_2 atmosphere, shelf stability, 1000 h (65%)	5.4
$FASnI_3$	SnF_2, FBH	20% RH, encapsulated, MPPT, 600 h (93%)	9.47
$FASnI_3$	$SnCl_2$, GA	20% RH, shelf stability, unencapsulated, 1000 h (80%)	9.03
$FASnI_3$	SnF_2, TFEACl	Continuous 1 sun irradiation, 350 h (60%)	5.3

$FASnI_3$	$SnCl_2$, KHQSA	20% RH, unencapsulated, continuous 1 sun irradiation, 16 h (50%)	6.76
$Cs_{0.2}FA_{0.8}SnI_3$	SnF_2, $SnCl_2$	N_2 atmosphere, encapsulated, MPPT, 1000 h (95%)	10.8
$MASnIBr_2$	SnF_2	N_2, shelf life, 60 days (80%	3.7
$FA_{0.75}MA_{0.25}SnI_3$	SnF_2, TM-DHP	N_2 atmosphere, shelf stability, unencapsulated, 50 d (100%)	11.5
$CsSnIBr_2$	SnF_2, HPA	20% RH, shelf stability, encapsulated, 77 d (100%)	3.2

3.3 Defect Parameter Characterization and Defect Tolerance

The search for potential candidates with exciting optoelectronic properties among lead-free perovskites needs a defect-tolerant absorber. It means that the defects present at low temperature on the semiconductor thin films are shallow with small capture cross section or it remains within the energy bands [94]. The promising materials of defect-tolerant absorbers are found to be large-sized metal cations with electronic configuration ns^2 and with high polarizability [94,95]. Theoretical calculations assisted in determining the material composition of suitable defect-tolerant absorbers [96,97]. However experimental validation is mandatory for the rational development of high-performing lead-free perovskite solar cells. In general, experimental defect tolerance is determined from space charge limited current characterization using total volumetric defect density [98,99]. A direct measure of defect tolerance from defect density, capture cross-section, and energy characteristics is still not followed. A rigorous experimental evaluation of defect densities surely helps identify potential lead-free perovskites and offers solutions for defect state passivation. Characterization of defect parameters through experimental approaches is of pivotal importance in developing lead-free perovskite solar cells.

3.4 Charge Transport Characterization

The photovoltaic performance of a solar cell essentially depends on the charge transport properties. Optimization of the photovoltaic performance of lead-free perovskite solar cells has special importance to its charge transport characteristics. Unfortunately, reports of experimental charge transport data are scanty and deficient. Consequently, hindering the progress of high-performing lead-free perovskite solar cells. Experimental characterization of charge transport is generally obtained from a single sweep space charge limited current (SSLC) approach. In most case, these data are inaccurate as it scans in a forward direction only and neglects any other important effects resulting because of the large applied field

[100]. To establish the accuracy of charge transport properties, a double-sweep method has to be adopted. Moreover, SSLC has to be validated against the expected thickness dependence of the absorber layer. Mobility of the charge carriers can also be determined from the DC-magnetic field Hall effect [101,102]. This effect can be successfully applied to determine the mobility of lead-free perovskites with high mobility and low resistivity for example tin-germanium and tin base absorbers. However, in the case of bismuth and antimony-based perovskites belonging to moderate/high resistivity leads to substantial error in charge transport data [87,103]. Thus, the Hall effect measurement data of the perovskite absorbers must be validated critically with the observed trend. Recently AC-magnetic field Hall effect has been demonstrated for low-mobility perovskite absorbers [104,105]. In the future, this approach can become an essential route for the characterization of charge transport properties of lead-free perovskites.

3.5   Electronic Dimensionality

Electronic dimensionality is one of the keys deciding factors in the performance potential of lead-free perovskite solar cells [59]. The charge transport properties and defect tolerance depend directly on electronic dimensionality but inversely on excitonic effects. Structural dimensionality and electronic dimensionality often overlap with each other as in 3-dimensional $ASnX_3$ and zero-dimensional Cs_2SnX_6. Similarly, bismuth and antimony-based $A_2B_3X_9$ systems in their dimer or layered phase have electronic dimensionality of 0.5D or 2.5D respectively. However, 3-dimensional cation-ordered double perovskite structures e.g. Cs_2AgBiX_6 are electronically zero-dimensional.

Leaving aside 3D $ASnX_3$ and $AGeX_3$ (stability issues), among lead-free perovskites bismuth and antimony-based $A_2B_3X_9$ systems belong to the highest electronic dimensionality of 2.5. The PCE values of $A_3Sb_2X_9$ and $A_3Bi_2X_9$ absorbers are similar or marginally smaller than the dimer phase and zero-dimensional CO-HDPs and VO-HDPs. The photovoltaic performance of the layered $A_2B_3X_9$ absorbers is dependent on the orientation of the planes of octahedra against the substrate. The exploitation of higher dimensional layered $A_2B_3X_9$ needs appropriate orientation either parallel or random orientation concerning the substrate. In this regard, suitable deposition methods play important role in enhancing their performance. Recently halide double perovskite structures, which are electronically 3D are proposed but their photovoltaic performance is not yet explored in detail. These materials are expected to have significant photovoltaic efficiency which needs special attention. Table 7 summarizes some important aspects of lead-free perovskites [106].

Table 7. Summary of lead-free perovskites and their derivatives [106]

Family	Stoichiometry	Chemical formula	Band Gap(eV)	PCE(%)	Synthesis Solution phase	Solid state
Perovskite	1-1-3	$FASnI_3$	1.41		Y	Y
		$CsSnI_3$	1.30		Y	Y
		$MASnI_3$	1.23	6.4	Y	
		$MASnI_3$	1.20-1.35		Y	Y
		$MASnI_{3-x}Br_x$	1.30-2.15	5.73		Y
		$MASnI_{1-x}Pb_xI_x$	1.21-1.54		Y	Y
VO-HDP	2-1-6	Cs_2SnCl_6	3.9	0.07	Y	
		Cs_2SnI_6	1.26	6.94	Y	
		Cs_2SnBr_6	2.7	0.04	Y	
		$Cs_2Sn_{1-x}Te_xI_6$	1.25-1.59		Y	
2D-perovskite	3-2-9	$MA_3Bi_2I_9$	2.1	0.12	Y	
		$MA_3Bi_2I_9$		0.2	Y	
		$MA_3Sb_2I_9$	2.14	0.5	Y	
		$Rb_3Sb_2I_9$	2.1-2.24	0.66	Y	
		$Cs_3Sb_2I_9$	2.05			Y
		$Cs_3Bi_2I_9$	2.2	1.09	Y	
HDP	2-1-1-6	$Cs_2BiAgCl_6$	2.2			Y
		$Cs_2BiAgCl_6$	2.77		Y	Y
		$Cs_2BiAgBr_6$	1.9		Y	Y
		$Cs_2BiAgBr_6$	1.95			Y
		$Cs_2BiAgBr_6$	2.19		Y	Y
		MA_2BiKCl_6	3.04		Y	
Others	1-2-7	$HDABiI_5$	2.1	0.027	Y	
Others	1-1-5	$AgBi_2I_7$	1.87	1.22	Y	

4. Processing of High-Quality Lead-Free Perovskite Films

The major task while improving the efficiencies of PSC devices is to parallelly improve the quality of film of the perovskite layer. The focus is to get a smooth, plain, uniform, and impurity-free layer. There are several techniques available to produce high-quality perovskite thin film. These are discussed in the following section.

4.1 Vapour deposition method

Vapor-deposition method is one of the earliest techniques among fabrication processes of PSCs initiated by Liu et al. Using this method, they demonstrated efficiency of more than 15% in a planar heterojunction PSC [107]. Numerous reports are available on the vapor deposition of SnI_2, $SnBr_2$, MAI, MABr, FAI, etc on a substrate. This method is fruitful in making and producing uniform films at a large scale and results in improved and better performance of the device. This method can also be utilized for fabricating multi-stack thin films with ease [108]. Xi et al. have prepared $FASnI3$ film from stacked layers of FAI/polymer by annealing followed by vaporization of SnI_2 [109]. Fabrication of stable $MASnBr_3$ film through co-evaporation as well as sequential evaporation and $CsSnBr_3$ film through the vapor deposition method has been recently reported [110, 111].

4.2 Anti-Solvent Technique

Another important method to produce good quality perovskite films is the anti-solvent method, also known as solvent engineering [Fig. 8a]. This method was reported by Jeon et al. who demonstrated the role of DMSO in slowing down the rate of crystal growth by forming an intermediate MAI- PbI_2- DMSO [112]. Since then this method has been proven to be of great use for fabrication and has also been endorsed by many researchers.

4.3 Solution Processing

The next widely used method is the solution fabrication method. This method has the advantage of low energy requirement even for large-scale manufacturing of films. However, due to anisotropic crystallization happening during the evaporation of the solvent, the production of a flat and compact film is of great challenge. Hence, it is necessary to control the rapid crystallization process during fabrication. Mostly, to produce high-quality perovskite films with excellent optoelectronic properties, kinetics between crystal grain growth and nucleation must be optimized.

(a)

(b)

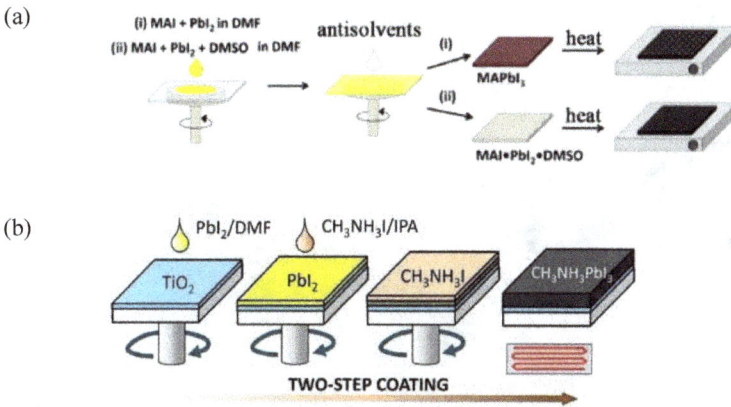

Figure 8. Fabrication methods of high-quality PSC films: (a) anti-solvents and [113] (b) two-step method [114]

4.4 Two-Step Deposition

The two-step deposition method can produce excellent-quality perovskite films [Fig. 8b]. This method was first reported by Burschka et al. [115]. He added PbI_2 to the porous TiO_2 film layer, and then it was exposed to MAI solution for processing the perovskite film. The film obtained by this method was better. Moreover, in a mesoporous device, the efficiency reached 15%. A two-step process of vapor deposition and the spin coating was utilized by Schlettwein et al. for the fabrication of $MASnI_3$ film [116]. The morphology of $MASnI_3$ was found to depend on the amount of MAI. Kanatzidis et al. improved the morphology of $MASnI_3$ film by controlling the growth using a low-temperature vapor-assisted solution process (LT-VASP) [Fig. 9a] [117].

4.5 Low Pressure Assisted Solution Processing

Tin-based PSCs are unstable due to the rapid oxidation of the Sn^{2+} to the Sn^{4+} state. This oxidation mechanism ruins the crystal structure and quality of the PSC film. Thus, the fabrication of tin perovskite films at ambient conditions is a great challenge. In this situation, low-pressure assisted solution processing helps to produce high-quality films under an ambient environment [Fig. 9b]. For example, Wu et al. have produced a high-quality film of $Cs_2AgBiBr_6$ using this method in an ambient environment [118]. Low pressure assisted solution processing technique has the advantage of producing high-density uniform perovskite films with single crystalline grains.

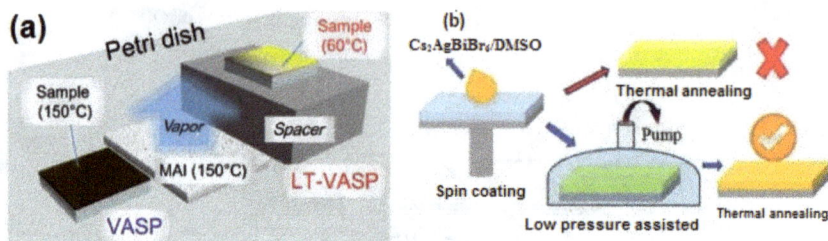

Figure 9. The fabrication method of PSC films by spin coating and evaporation (a) vapor-assisted solution process (VASP) and low-temperature VASP (LT-VASP)processes [117]. (b) Low-pressure assisted solution processing [118].

4.6   Spin Coating

Spin coating [Fig.10] is the most routinely used method for preparing PSC films [119]. It is performed inside a glove box in a nitrogen atmosphere to protect tin-based PSC material from moisture and oxygen. The spin coating process is not beneficial for the large-scale production of PSC films as this method leads to a large amount of material waste [120].

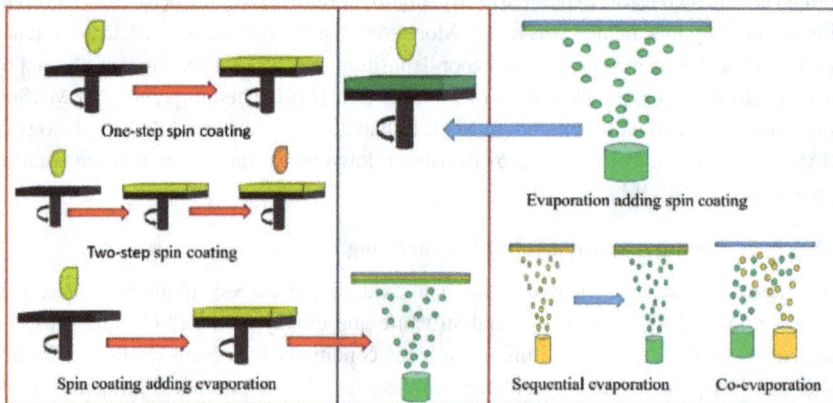

Figure 10. The fabrication method of PSC films by spin coating and evaporation [58]

4.7   Inter-diffusion Method

Moving forward, the next method that was developed is known as the inter-diffusion method initiated by Xiao et al.[121] In a planar device, sequential spin coating of MAI and PbI_2 improved the power conversion efficiency to 15.4%. The efficiency can be further enhanced by treating with DMF solvent annealing. In a very recent study, it was shown that PMMA can control the crystal growth and nucleation of the films leading to larger grains and with fewer defects [122]. Thus PSCs receive a huge leap in efficiency to reach a 21.02% level.

4.8   Doctor Blade Coating

To fabricate pinhole-free high-quality PSC films on a large area, doctor blade coating has been developed [Fig.11a]. This method does not require rigorous control of temperature and precursor stoichiometry. This method can be thought of as a complementary method of the spin coating technique. Deng et al. prepared $MA_{0.6}FA_{0.4}PbI_3$ film with a power conversion efficiency of 18.3% using this method [123].

4.9   Vacuum Flash-Assisted Solution Process (VASP)

Vacuum Flash Assisted Solution Process (VASP) is relatively a new method for the fabrication of highly efficient perovskite films for large-area applications [Fig.11b]. Using this method high quality crystals with a power conversion efficiency of 19.6% were obtained [124].

4.10   Complex Assisted Gas Quenching (CAGQ) method

The CAGQ method is developed by Connins et al [Fig.11c]. In this process, an intermediate DMSO complex is formed while the residue solvent is removed by flowing nitrogen gas. Using this method efficiency obtained was up to 18% [125].

4.11   Soft Cover Deposition (SCD)

SCD method produces uniform, large-grain PSC films of a large area. It was developed by Ye et al. while fabricating $MAPbI_3$ films [126]. The edge of this event is that it can improve the utilization efficiency of the material by up to 80% which is much higher compared to the spin coating technique giving 1% only. The power conversion efficiency obtained for a 1 cm^2 area with PSCs films by this method was 17.6%.

Conclusion and outlook

In this chapter current status and progress of environmentally safe lead-free perovskite solar cells (PSCs) have been discussed. Lead-free perovskites help in removing lead toxicity from the environment and may be pushed toward commercial applications. Perovskite absorbers based on isovalent and heterovalent substitution of lead with Sn(II), Cu(II), Bi(III), Sb(III), Sn(IV), Ti(IV) have been presented which are selected based on the perspective of stability and device performance. Despite several advantages of tin and germanium-based PSCs, they face the issue of stability due to oxidation. To date, the performance of bismuth and antimony-based PSCs has not reached the desired level. The efficiency of copper-based PSCs is low due to their wide band gap and weak charge transport behavior. Among different types of perovskites, halide double perovskites have a high potential for commercial development for their tunable band gap and great stability.

Various possible issues that are present in their developmental prospect have been put up in detail. Especially the effects of dimensionality, additives, and charge transport properties on the device performance are analyzed. The performance of lead-free PSCs can be improved by modulating band gaps, modifying film morphology, structural change, altering synthetic methods, etc. Computational simulation work can facilitate realizing the fundamentals of improving stability, carrier transport properties, and efficiency of perovskite materials. Combined efforts of simulation and experimental research work can soon bring breakthrough results in lead-free PSCs. The development of sophisticated fabrication techniques can lift the efficiency of PSC devices to theoretically predict values by improving the film quality. However, for real applications on a large scale, continuous efforts are necessary to produce high-quality PSC films. Overall integrating innovative perovskite materials for device fabrication and scale-up strategy is of top priority for the advancement of lead-free PSCs to solve the problem of affordable and clean energy.

References

[1] A.K. Jena, A. Kulkarni, T. Miyasaka, Halide perovskite photovoltaics: Background, status, and future prospects, Chem. Rev. 119 (2019) 3036-3103. https://doi.org/10.1021/acs.chemrev.8b00539

[2] P.K. Nayak, S. Mahesh, H.J. Snaith, D. Cahen, Photovoltaic solar cell technologies: Analyzing the state of the art, Nat. Rev. Mater. 4 (2019) 269-285. https://doi.org/10.1038/s41578-019-0097-0

[3] Y.H. Kim, J.S. Kim, T.W. Lee, Perovskite LEDs: Strategies to improve luminescence efficiency of metal-halide perovskites and light-emitting diodes, Adv. Mater. 31 (2019) 1970335. https://doi.org/10.1002/adma.201970335

[4] S.P. Senanayak, M.A. Jalebi, V.S. Kamboj, R. Carey, R. Shivanna, T. Tian, G. Schweicher, J. Wang, N. Giesbrecht, D.D. Nuzzo, H.E. Beere, P. Docampo, D.A. Ritchie, D. F. Jimenez, R.H. Friend, H. Sirringhaus, A general approach for hysteresis-

free, operationally stable metal halide perovskite field-effect transistors, Sci. Adv. 6 (2020) eaaz4948. https://doi.org/10.1126/sciadv.aaz4948

[5] V. Pecunia, Efficiency and spectral performance of narrowband organic and perovskite photodetectors: A cross-sectional review, J. Phys. Mater. 2 (2019) 042001. https://doi.org/10.1088/2515-7639/ab336a

[6] S.V.N. Pammi, R. Maddaka, V.D. Tran, J.H. Eom, V. Pecunia, S. Majumder, M.D. Kim, S.G. Yoon, CVD-deposited hybrid lead halide perovskite films for high-responsivity, self-powered photodetectors with enhanced photo stability under ambient conditions, Nano Energy 74 (2020) 104872. https://doi.org/10.1016/j.nanoen.2020.104872

[7] National Renewable Energy Laboratory, 2020. https://www.nrel.gov/pv/cell

[8] B. Murali, S. Dey, A.L. Abdelhady, W. Peng, E. Alarousu, A.R. Kirmani, N. Cho, S.P. Sarmah, M.R. Parida, M.I. Saidaminov, A.A. Zhumekenov, J. Sun, M.S. Alias, E. Yengel, B.S. Ooi, A. Amassian, O.M. Bakr, O.F. Mohammed, Surface restructuring of hybrid perovskite crystals, ACS Energy Lett. 1 (2016) 1119-1126. https://doi.org/10.1021/acsenergylett.6b00517

[9] J.A. Christians, P.A.M. Herrera, P.V. Kamat, Transformation of the excited state and photovoltaic efficiency of CH3NH3PbI3 perovskite upon controlled exposure to humidified air, J. Am. Chem. Soc. 137 (2015) 1530-1538. https://doi.org/10.1021/ja511132a

[10] W. Huang, J.S. Manser, P.V. Kamat, S. Ptasinska, Evolution of chemical composition, morphology, and photovoltaic efficiency of CH3NH3PbI3 perovskite under ambient conditions, Chem. Mater. 28 (2016) 303-311. https://doi.org/10.1021/acs.chemmater.5b04122

[11] W. Peng, X. Miao, V. Adinolfi, E. Alarousu, O.E. Tall, A.H. Emwas, C. Zhao, G. Walters, J. Liu, O. Ouellette, J. Pan, B. Murali, E.H. Sargent, O.F. Mohammed, O.M. Bakr, Engineering of CH3NH3PbI3 perovskite crystals by alloying large organic cations for enhanced thermal stability and transport properties, Angew. Chem. Int. Ed. 55 (2016) 10686-10690. https://doi.org/10.1002/anie.201604880

[12] F. Ma, J. Li, W. Li, N. Lin, L. Wang, J. Qiao, Stable α/δ phase junction of Formamidinium Lead Iodide perovskites for enhanced near-infrared emission, Chem. Sci. 8 (2017) 800-805. https://doi.org/10.1039/C6SC03542F

[13] G.A. Tosado, Y.Y. Lin, E. Zheng, Q. Yu, Impact of Cesium on the phase and device stability of triple cation Pb-Sn double halide perovskite films and solar cells, J. Mater. Chem. 6 (2018) 17426-17436. https://doi.org/10.1039/C8TA06391E

[14] A. Babayigit, A. Ethirajan, M. Muller, B. Conings, Toxicity of organometal halide perovskite solar cells, Nat. Mater. 15 (2016) 247-251. https://doi.org/10.1038/nmat4572

[15] I.R. Benmessaoud, A.L.M.-Mellier, E. Horváth, B. Maco, M. Spina, H.A. Lashuel, L. Forró, Health hazards of Methylammonium Lead Iodide based perovskites: Cytotoxicity studies, Toxicol. Res. 5 (2016) 407-419. https://doi.org/10.1039/C5TX00303B

[16] A. Babayigit, D.D. Thanh, A. Ethirajan, J. Manca, M. Muller, H.-G. Boyen, B. Conings, Assessing the toxicity of Pb-and Sn-based perovskite solar cells in model organism Danio rerio, Sci. Rep. 6 (2016) 1-11. https://doi.org/10.1038/srep18721

[17] R.Wang, M. Mujahid, Y. Duan, Z.K. Wang, J. Xue, Y. Yang, A review of perovskites solar cell stability, Adv. Funct. Mater. 29 (2019) 1808843. https://doi.org/10.1002/adfm.201808843

[18] A. Abate, Perovskite solar cells go lead free, Joule.1 (2017) 659-664. https://doi.org/10.1016/j.joule.2017.09.007

[19] W.F. Yang, F. Igbari, Y.H. Lou, Z.K. Wang, L.S. Liao, Tin halide perovskites: Progress and challenges, Adv. Energy Mater. 10 (2020) 1902584. https://doi.org/10.1002/aenm.201902584

[20] K. Nishimura, M.A. Kamarudin, D. Hirotani, K. Hamada, Q. Shen, S. Iikubo, T. Minemoto, K. Yoshino, S. Hayase, Lead-free Tin-halide perovskite solar cells with 13% efficiency, Nano Energy. 74 (2020) 104858. https://doi.org/10.1016/j.nanoen.2020.104858

[21] W. Ke, C.C. Stoumpos, M.G. Kanatzidis, Unleaded perovskites: Status quo and future prospects of Tin-based perovskite solar cells, Adv. Mater. 31 (2019) 1803230. https://doi.org/10.1002/adma.201803230

[22] P. Xu, S. Chen, H.J. Xiang, X.G. Gong, S.H. Wei, Influence of defects and synthesis conditions on the photovoltaic performance of perovskite semiconductor CsSnI3, Chem. Mater. 26 (2014) 6068-6072. https://doi.org/10.1021/cm503122j

[23] G. Xing, M.H. Kumar, W.K. Chong, X. Liu, Y. Cai, H. Ding, M. Asta, M. Grätzel, S. Mhaisalkar, N. Mathews, T.C. Sum, Solution-processed Tin-based perovskite for

near-infrared lasing, Adv. Mater. 28 (2016) 8191-8196.
https://doi.org/10.1002/adma.201601418

[24] M. Leng, Y. Yang, K. Zeng, Z. Chen, Z. Tan, S. Li, J. Li, B. Xu, D. Li, M.P. Hautzinger, Y. Fu, T. Zhai, L. Xu, G. Niu, S. Jin, J. Tang, All-inorganic Bismuth-based perovskite quantum dots with bright blue photoluminescence and excellent stability, Adv. Funct. Mater. 28 (2018) 1704446. https://doi.org/10.1002/adfm.201704446

[25] M. Lyu, J.H. Yun, P. Chen, M. Hao, L. Wang, Addressing toxicity of Lead: Progress and applications of low-toxic metal halide perovskites and their derivatives, Adv. Energy Mater. 7 (2017) 1602512. https://doi.org/10.1002/aenm.201602512

[26] A.B.F. Vitoreti, S. Agouram, M.S.D.L. Fuente, V.M. Sanjosé, M.A. Schiavon, I.M. Seró, Study of the partial substitution of Pb by Sn in Cs-Pb-Sn-Br nanocrystals owing to obtaining stable nanoparticles with excellent optical properties, J. Phys. Chem: C, 122 (2018) 14222-14231. https://doi.org/10.1021/acs.jpcc.8b02499

[27] D. Bartesaghi, A. Ray, J. Jiang, R.K.M. Bouwer, S. Tao, T.J. Savenije, Partially replacing Pb2+ by Mn2+ in hybrid metal halide perovskites: Structural and electronic properties, APL Mater. 6 (2018) 121106. https://doi.org/10.1063/1.5060953

[28] J. Liu, G. Wang, Z. Song, X. He, K. Luo, Q. Ye, C. Liao, J. Mei, FAPb− xSnxI3 mixed metal halide perovskites with improved light harvesting and stability for efficient planar heterojunction solar cells, J. Mater. Chem: A, 5 (2017) 9097-9106. https://doi.org/10.1039/C6TA11181E

[29] B. Zhao, M.A. Jalebi, M. Tabachnyk, H. Glass, V.S. Kamboj, W. Nie, A.J. Pearson, Y. Puttisong, K.C. Gödel, H. E. Beere, D.A. Ritchie, A.D. Mohite, S.E. Dutton, R.H. Friend, A. Sadhanala, High open-circuit voltages in Tin-rich low-bandgap perovskite-based planar heterojunction photovoltaics, Adv. Mater. 29 (2017) 1604744. https://doi.org/10.1002/adma.201604744

[30] F. Yang, D. Hirotani, G. Kapil, M.A. Kamarudin, C.H. Ng, Y. Zhang, Q. Shen, S. Hayase, All-inorganic CsPb1− xGexI2Br perovskite with enhanced phase stability and photovoltaic performance, Angew. Chem. Int. Ed. 130 (2018) 12927-12931. https://doi.org/10.1002/ange.201807270

[31] M.R. Filip, F. Giustino, Computational screening of homovalent Lead substitution in organic-inorganic halide perovskites, J. Phys. Chem: C, 120 (2016) 166-173. https://doi.org/10.1021/acs.jpcc.5b11845

[32] S.F. Hoefler, G. Trimmel, T. Rath, Progress on Lead-free metal halide perovskites for photovoltaic applications: A review, Monatsh. Chem. 148 (2017) 795-826. https://doi.org/10.1007/s00706-017-1933-9

[33] L. Yang, A.T. Barrows, D.G. Lidzey, T. Wang, Recent progress and challenges of organometal halide perovskite solar cells, Rep. Prog. Phys. 79 (2016) 026501. https://doi.org/10.1088/0034-4885/79/2/026501

[34] G. Nagabhushana, R. Shivaramaiah, A. Navrotsky, Direct calorimetric verification of thermodynamic instability of Lead halide hybrid perovskites, Proc. Natl. Acad. Sci. 113 (2016) 7717-7721. https://doi.org/10.1073/pnas.1607850113

[35] B. Saparov, D.B. Mitzi, Organic-inorganic perovskites: Structural versatility for functional materials design, Chem. Rev. 116 (2016) 4558-4596. https://doi.org/10.1021/acs.chemrev.5b00715

[36] T. Krishnamoorthy, H. Ding, C. Yan, W.L. Leong, T. Baikie, Z. Zhang, M. Sherburne, S. Li, M. Asta, N. Mathews, S.G. Mhaisalkar, Lead-free Germanium Iodide perovskite materials for photovoltaic applications, J. Mater. Chem: A, 3 (2015) 23829-23832. https://doi.org/10.1039/C5TA05741H

[37] W.F. Yang, F. Igbari, Y.H. Lou, Z.K. Wang, L.S. Liao, Tin halide perovskites: Progress and challenges, Adv. Energy Mater. 10 (2020) 1902584. https://doi.org/10.1002/aenm.201902584

[38] K. Nishimura, M.A. Kamarudin, D. Hirotani, K. Hamada, Q. Shen, S. Iikubo, T. Minemoto, K. Yoshino, S. Hayase, Lead-free Tin-halide perovskite solar cells with 13% efficiency, Nano Energy. 74 (2020) 104858. https://doi.org/10.1016/j.nanoen.2020.104858

[39] W. Ke, C.C. Stoumpos, M.G. Kanatzidis, Unleaded perovskites: Status quo and future prospects of Tin-based perovskite solar cells, Adv. Mater. 31 (2019) 1803230. https://doi.org/10.1002/adma.201803230

[40] W. Ke, M.G. Kanatzidis, Prospects for low-toxicity Lead-free perovskite solar cells, Nat. Commun. 10 (2019) 1-4. https://doi.org/10.1038/s41467-018-07882-8

[41] M.E. Kayesh, T.H. Chowdhury, K. Matsuishi, R. Kaneko, S. Kazaoui, J.J. Lee, T. Noda, A. Islam, Enhanced photovoltaic performance of FASnI3-based perovskite solar cells with Hydrazinium Chloride coadditive, ACS Energy Lett. 3 (2018) 1584-1589. https://doi.org/10.1021/acsenergylett.8b00645

[42] Q. Tai, X. Guo, G. Tang, P. You, T.W. Ng, D. Shen, J. Cao, C.K. Liu, N. Wang, Y. Zhu, C.S. Lee, F. Yan, Antioxidant grain passivation for air-stable Tin-based

perovskite solar cells, Angew. Chem. Int. Ed. 58 (2019) 806-810. https://doi.org/10.1002/anie.201811539

[43] X. Meng, T. Wu, X. Liu, X. He, T. Noda, Y. Wang, H. Segawa, L. Han, Highly reproducible and efficient FASnI3 perovskite solar cells fabricated with volatilizable reducing solvent, J. Phys. Chem. Lett. 11 (2020) 2965-2971. https://doi.org/10.1021/acs.jpclett.0c00923

[44] T. Wang, Q. Tai, X. Guo, J. Cao, C.K. Liu, N. Wang, D. Shen, Y. Zhu, C.S. Lee, F. Yan, Highly air-stable Tin-based perovskite solar cells through grain-surface protection by Gallic acid, ACS Energy Lett. 5 (2020) 1741-1749. https://doi.org/10.1021/acsenergylett.0c00526

[45] E. Jokar, C.H. Chien, C.M. Tsai, A. Fathi, EW.G. Diau, Robust Tin-based perovskite solar cells with hybrid organic cations to attain efficiency approaching 10%, Adv. Mater. 31 (2019) 1804835. https://doi.org/10.1002/adma.201804835

[46] M.A. Kamarudin, D. Hirotani, Z. Wang, K. Hamada, K. Nishimura, Q. Shen, T. Toyoda, S. Iikubo, T. Minemoto, K. Yoshino, S. Hayase, Suppression of charge carrier recombination in Lead-free Tin halide perovskite via Lewis base post-treatment, J. Phys. Chem. Lett. 10 (2019) 5277-5283. https://doi.org/10.1021/acs.jpclett.9b02024

[47] W. Ke, C.C. Stoumpos, I. Spanopoulos, L. Mao, M. Chen, M.R. Wasielewski, M.G. Kanatzidis, Efficient Lead-free solar cells based on hollow {en} MASnI3 perovskites, J. Am. Chem. Soc. 139 (2017) 14800-14806. https://doi.org/10.1021/jacs.7b09018

[48] W. Ke, C.C. Stoumpos, M. Zhu, L. Mao, I. Spanopoulos, J. Liu, O.Y. Kontsevoi, M. Chen, D. Sarma, Y. Zhang, M.R. Wasielewski, M.G. Kanatzidis, Enhanced photovoltaic performance and stability with a new type of hollow 3D perovskite {en} FASnI3, Sci. Adv. 3 (2017) 1701293. https://doi.org/10.1126/sciadv.1701293

[49] S. Shao, J. Liu, G. Portale, H.H. Fang, G.R. Blake, G.H.T. Brink, L.J.A. Koster, M.A. Loi, Highly reproducible Sn-based hybrid perovskite solar cells with 9% efficiency, Adv. Energy Mater. 8 (2018) 1702019. https://doi.org/10.1002/aenm.201702019

[50] X. Jiang, F. Wang, Q. Wei, H. Li, Y. Shang, W. Zhou, C. Wang, P. Cheng, Q. Chen, L. Chen, Z. Ning, Ultra-high open-circuit voltage of Tin perovskite solar cells via an electron transporting layer design, Nat. Commun. 11 (2020) 1-7. https://doi.org/10.1038/s41467-019-13993-7

[51] F. Wang, X. Jiang, H. Chen, Y. Shang, H. Liu, J. Wei, W. Zhou, H. He, W. Liu, Z. Ning, 2D-quasi-2D-3D hierarchy structure for Tin perovskite solar cells with enhanced

efficiency and stability, Joule. 2 (2018) 2732-2743.
https://doi.org/10.1016/j.joule.2018.09.012

[52] T. Wu, X. Liu, X. Luo, X. Lin, D. Cui, Y. Wang, H. Segawa, Y. Zhang, L. Han,
Lead-free Tin perovskite solar cells, Joule. 5 (2021) 863-886.
https://doi.org/10.1016/j.joule.2021.03.001

[53] T. Krishnamoorthy, H. Ding, C. Yan, W.L. Leong, T. Baikie, Z. Zhang, M.
Sherburne, S. Li, M. Asta, N. Mathews, S.G. Mhaisalkar, Lead-free Germanium
Iodide perovskite materials for photovoltaic applications, J. Mater. Chem: A, 3 (2015)
23829- 23832. https://doi.org/10.1039/C5TA05741H

[54] I. Kopacic, B. Friesenbichler, S.F. Hoefler, B. Kunert, H. Plank, T. Rath, G.
Trimmel, Enhanced performance of Germanium halide perovskite solar cells through
compositional engineering, ACS Appl. Energy Mater. 1 (2018) 343-347.
https://doi.org/10.1021/acsaem.8b00007

[55] V. Pecunia, L.G. Occhipinti, A. Chakraborty, Y. Pan, Y. Pang, Lead-free halide
perovskite photovoltaics: Challenges, open questions, and opportunities, APL Mater. 8
(2020) 100901. https://doi.org/10.1063/5.0022271

[56] C.H. Ng, K. Nishimura, N. Ito, K. Hamada, D. Hirotani, Z. Wang, F. Yang, S.
Likubo, Q. Shen, K. Yoshino, T. Minemoto, S. Hayase, Role of GeI2 and SnF2
additives for SnGe perovskite solar cells, Nano Energy. 58 (2019) 130-137.
https://doi.org/10.1016/j.nanoen.2019.01.026

[57] M. Chen, M.G. Ju, H.F. Garces, A.D. Carl, L.K. Ono, Z. Hawash, Y. Zhang, T.
Shen, Y. Qi, R.L. Grimm, D. Pacifici, X.C. Zeng, Y. Zhou, N.P. Padture, Highly stable
and efficient all-inorganic Lead-free perovskite solar cells with native-oxide
passivation, Nat. Commun. 10 (2019) 1-8. https://doi.org/10.1038/s41467-018-07882-
8

[58] R. Wang, J. Wang, S. Tan, Y. Duan, Z.-K. Wang, Y. Yang, Opportunities and
challenges of Lead-free perovskite optoelectronic devices, Trends Chem. 1 (2019)
368-379. https://doi.org/10.1016/j.trechm.2019.04.004

[59] Z. Xiao, W. Meng, J. Wang, D.B. Mitzi, Y. Yan, Searching for promising new
perovskite-based photovoltaic absorbers: The importance of electronic dimensionality,
Mater. Horiz. 4 (2017) 206-216. https://doi.org/10.1039/C6MH00519E

[60] H. Hu, B. Dong, W. Zhang, Low-toxic metal halide perovskites: Opportunities and
future challenges, J. Mater. Chem: A, 5 (2017) 11436-11449.
https://doi.org/10.1039/C7TA00269F

[61] Park, B. Wook, B. Philippe, X. Zhang, H. Rensmo, G. Boschloo, E.M.J. Johansson, Bismuth based hybrid perovskites A3Bi2I9 (A: Methylammonium or Cesium) for Solar cell application, Adv. Mater. 27 (2015) 6806-6813. https://doi.org/10.1002/adma.201501978

[62] B.W. Park, B. Philippe, X. Zhang, H. Rensmo, G. Boschloo, E.M.J. Johansson, Bismuth based hybrid perovskites A3Bi2I9 (A: methylammonium or cesium) for solar cell application, Adv. Mater. 27 (2015) 6806-6813. https://doi.org/10.1002/adma.201501978

[63] J.P.C. Baena, L. Nienhaus, R.C. Kurchin, S.S. Shin, S. Wieghold, N.T.P. Hartono, M. Layurova, N.D. Klein, J.R. Poindexter, A. Polizzotti, S. Sun, M.G. Bawendi, T. Buonassisi, A-Site cation in inorganic A3Sb2I9 perovskite influences structural dimensionality, exciton binding energy, and solar cell performance, Chem. Mater. 30 (2018) 3734- 3742. https://doi.org/10.1021/acs.chemmater.8b00676

[64] J.C. Hebig, I. Kuhn, J. Flohre, T. Kirchartz, Optoelectronic properties of (CH3NH3)3Sb2I9 thin films for photovoltaic applications, ACS Energy Lett. 1 (2016) 309-314. https://doi.org/10.1021/acsenergylett.6b00170

[65] S.M. Jain, D. Phuyal, M.L. Davies, M. Li, B. Philippe, C.D. Castro, Z. Qiu, J. Kim, T. Watson, W.C. Tsoi, O. Karis, H. Rensmo, G. Boschloo, T. Edvinsson, J.R. Durrant, An effective approach of vapor assisted morphological tailoring for reducing metal defect sites in Lead-free, (CH3NH3)3Bi2I9 Bismuth-based perovskite solar cells for improved performance and long-term stability, Nano Energy. 49 (2018) 614-624. https://doi.org/10.1016/j.nanoen.2018.05.003

[66] F. Bai, Y. Hu, Y. Hu, T. Qiu, X. Miao, S. Zhang, Lead-free, air-stable ultrathin Cs3Bi2I9 perovskite nanosheets for solar cells, Sol. Energy Mater. Sol. Cells. 184 (2018) 15-21. https://doi.org/10.1016/j.solmat.2018.04.032

[67] S. Weber, T. Rath, K. Fellner, R. Fischer, R. Resel, B. Kunert, T. Dimopoulos, A. Steinegger, G. Trimmel, Influence of the Iodide to Bromide ratio on crystallographic and optoelectronic properties of Rubidium Antimony halide perovskites, ACS Appl. Energy Mater. 2 (2019) 539-547. https://doi.org/10.1021/acsaem.8b01572

[68] F. Li, Y. Wang, K. Xia, R.L.Z. Hoye, V. Pecunia, Microstructural and photoconversion efficiency enhancement of compact films of Lead-free perovskite derivative Rb3Sb2I9, J. Mater. Chem: A, 8 (2020) 4396-4406. https://doi.org/10.1039/C9TA13352F

[69] Y. Yang, C. Liu, M. Cai, Y. Liao, Y. Ding, S. Ma, X. Liu, M. Guli, S. Dai, M.K. Nazeeruddin, Dimension-controlled growth of Antimony-based perovskite-like halides

for Lead-free and semitransparent photovoltaics, ACS Appl. Mater. Interfaces. 12 (2020) 17062-17069. https://doi.org/10.1021/acsami.0c00681

[70] Y. Peng, F. Li, Y. Wang, Y. Li, R.L.Z. Hoye, L. Feng, K. Xia, V. Pecunia, Enhanced photoconversion efficiency in Cesium-Antimony-halide perovskite derivatives by tuning crystallographic dimensionality, Appl. Mater. Today. 19 (2020) 100637. https://doi.org/10.1016/j.apmt.2020.100637

[71] M. Wang, W. Wang, B. Ma, W. Shen, L. Liu, K. Cao, S. Chen, W. Huang, Lead free perovskite materials for solar cells, Nano-Micro Lett. 13 (2021) 1-36. https://doi.org/10.1007/s40820-020-00525-y

[72] F. Igbari, Z.K. Wang, L.S. Liao, Progress of Lead-free halide double perovskites, Adv. Energy Mater. 9 (2019) 1803150. https://doi.org/10.1002/aenm.201803150

[73] P.K. Kung, M.H. Li, P.Y. Lin, J.Y. Jhang, M. Pantaler, D.C. Lupascu, G. Grancini, P. Chen, Lead-free double perovskites for perovskite solar cells, Sol. RRL. 4 (2020) 1900306. https://doi.org/10.1002/solr.201900306

[74] X.G. Zhao, D. Yang, J.C. Ren, Y. Sun, Z. Xiao, L. Zhang, Rational design of halide double perovskites for optoelectronic applications, Joule. 2 (2018) 1662-1673. https://doi.org/10.1016/j.joule.2018.06.017

[75] I.N. Flerov, M.V. Gorev, K.S. Aleksandrov, A. Tressaud, J. Grannec, M. Couzi, Phase transitions in elpasolites (ordered perovskites), Mater. Sci. Eng: R, 24 (2018) 81-151. https://doi.org/10.1016/S0927-796X(98)00015-1

[76] R. Kentsch, M. Scholz, J. Horn, D. Schlettwein, K. Oum, T. Lenzer, Exciton dynamics and electron-phonon coupling affect the photovoltaic performance of the Cs2AgBiBr6 double perovskite, J. Phys. Chem: C, 122 (2018) 25940-25947. https://doi.org/10.1021/acs.jpcc.8b09911

[77] C.N. Savory, A. Walsh, D.O. Scanlon, Can Pb-free halide double perovskites support high-efficiency solar cells?, ACS Energy Lett. 1 (2016) 949-955. https://doi.org/10.1021/acsenergylett.6b00471

[78] A.H. Slavney, T. Hu, A.M. Lindenberg, H.I. Karunadasa, A Bismuth-halide double perovskite with long carrier recombination lifetime for photovoltaic applications, J. Am. Chem. Soc. 138 (2016) 2138-2141. https://doi.org/10.1021/jacs.5b13294

[79] M.R. Filip, X. Liu, A. Miglio, G. Hautier, F. Giustino, Phase diagrams and stability of Lead-free halide double perovskites Cs2BB′ X6: B= Sb and Bi, B′= Cu, Ag, and Au, and X= Cl, Br, and I, J. Phys. Chem: C, 122 (2018) 158-170. https://doi.org/10.1021/acs.jpcc.7b10370

[80] S.E. Creutz, E.N. Crites, M.C.D. Siena, D.R. Gamelin, Colloidal nanocrystals of Lead-free double-perovskite (elpasolite) semiconductors: Synthesis and anion exchange to access new materials, Nano Lett. 18 (2018) 1118-1123. https://doi.org/10.1021/acs.nanolett.7b04659

[81] X. Qiu, B. Cao, S. Yuan, X. Chen, Z. Qiu, Y. Jiang, Q. Ye, H. Wang, H. Zeng, J. Liu, M.G. Kanatzidis, From unstable CsSnI3 to air-stable Cs2SnI6: A Lead-free perovskite solar cell light absorber with bandgap of 1.48 eV and high absorption coefficient, Sol. Energy Mater. Sol. Cells. 159 (2017) 227-234. https://doi.org/10.1016/j.solmat.2016.09.022

[82] T. Kirchartz, U. Rau, What makes a good solar cell?, Adv. Energy Mater. 8 (2018) 1703385. https://doi.org/10.1002/aenm.201703385

[83] D. Sabba, H.K. Mulmudi, R.R. Prabhakar, T. Krishnamoorthy, T. Baikie, P.P. Boix, S. Mhaisalkar, N. Mathews, Impact of anionic Br-substitution on open circuit voltage in Lead free perovskite (CsSnI3-xBr x) solar cells, J. Phys. Chem: C, 119 (2015) 1763-1767. https://doi.org/10.1021/jp5126624

[84] W. Ming, H. Shi, M.H. Du, Large dielectric constant, high acceptor density, and deep electron traps in perovskite solar cell material CsGe3, J. Mater. Chem: A, 4 (2016) 13852-13858. https://doi.org/10.1039/C6TA04685A

[85] F. Li, Y. Wang, K. Xia, R.L.Z. Hoye, V. Pecunia, Microstructural and photoconversion efficiency enhancement of compact films of Lead-free perovskite derivative Rb3Sb2I9, J. Mater. Chem: A, 8 (2020) 4396- 4406. https://doi.org/10.1039/C9TA13352F

[86] B. Saparov, F. Hong, J.P. Sun, H.S. Duan, W. Meng, S. Cameron, I.G. Hill, Y. Yan, D.B. Mitzi, Thin-film preparation and characterization of Cs3Sb2I9 : A Lead-free layered perovskite semiconductor, Chem. Mater. 27 (2015) 5622- 5632. https://doi.org/10.1021/acs.chemmater.5b01989

[87] P.C. Harikesh, H.K. Mulmudi, B. Ghosh, T.W. Goh, Y.T. Teng, K. Thirumal, M. Lockrey, K. Weber, T.M. Koh, S. Li, S. Mhaisalkar, N. Mathews, Rb as an alternative cation for templating inorganic Lead-free perovskites for solution processed photovoltaics, Chem. Mater. 28 (2016) 7496-7504. https://doi.org/10.1021/acs.chemmater.6b03310

[88] W. Meng, X. Wang, Z. Xiao, J. Wang, D.B. Mitzi, Y. Yan, Parity-forbidden transitions and their impact on the optical absorption properties of Lead-free metal halide perovskites and double perovskites, J. Phys. Chem. Lett. 8 (2017) 2999-3007. https://doi.org/10.1021/acs.jpclett.7b01042

[89] X.G. Zhao, D. Yang, J.C. Ren, Y. Sun, Z. Xiao, L. Zhang, Rational design of halide double perovskites for optoelectronic applications, Joule. 2 (2018) 1662-1673. https://doi.org/10.1016/j.joule.2018.06.017

[90] W. Gao, C. Ran, J. Xi, B. Jiao, W. Zhang, M. Wu, X. Hou, Z. Wu, High-quality Cs2AgBiBr6 double perovskite film for Lead-free inverted planar heterojunction solar cells with 2.2% efficiency, ChemPhysChem. 19 (2018) 1696-1700. https://doi.org/10.1002/cphc.201800346

[91] B. Lee, A. Krenselewski, S.I. Baik, D.N. Seidman, R.P.H. Chang, Solution processing of air-stable molecular semiconducting Iodosalts, Cs2SnI6−xBrx, for potential solar cell applications, Sustain. Energy Fuels. 1 (2017) 710-724. https://doi.org/10.1039/C7SE00100B

[92] M.V. Khenkin, E.A. Katz, A. Abate, G. Bardizza, J.J. Berry, C. Brabec, F. Brunetti, V. Bulovi'c, Q. Burlingame, A.D. Carlo, R. Cheacharoen, Y.-B. Cheng, A. Colsmann, S. Cros, K. Domanski, M. Dusza, C.J. Fell, S.R. Forrest, Y. Galagan, D.D. Girolamo, M. Gratzel, A. Hagfeldt, E.V. Hauff, H. Hoppe, J. Kettle, H. Kobler, M.S. Leite, S. Liu, Y.-L. Loo, J.M. Luther, C.-Q. Ma, M. Madsen, M. Manceau, M. Matheron, M. McGehee, R. Meitzner, M.K. Nazeeruddin, A.F. Nogueira, C. Odaba, A. Osherov, N.-G. Park, M.O. Reese, F.D. Rossi, M. Saliba, U.S. Schubert, H. J. Snaith, S.D. Stranks, W. Tress, P.A. Troshin, V. Turkovic, S. Veenstra, I.V. Fisher, A. Walsh, T. Watson, H. Xie, R. Yildirim, S.M. Zakeeruddin, K. Zhu, M.L. Cantu, Consensus statement for stability assessment and reporting for perovskite photovoltaics based on ISOS procedures, Nat. Energy. 5 (2020) 35-49. https://doi.org/10.1038/s41560-019-0529-5

[93] J. Cao, F. Yan, Recent progress in Tin-based perovskite solar cells, Energy Environ. Sci. 14 (2021) 1286. https://doi.org/10.1039/D0EE04007J

[94] R.E. Brandt, J.R. Poindexter, P. Gorai, R.C. Kurchin, R.L.Z. Hoye, L. Nienhaus, M.W.B. Wilson, J.A. Polizzotti, R. Sereika, R. Žaltauskas, L.C. Lee, J.L.M.M. Driscoll, M. Bawendi, V. Stevanovic, T. Buonassisi, Searching for "defect-tolerant" photovoltaic materials: Combined theoretical and experimental screening, Chem. Mater. 29 (2017) 4667-4674. https://doi.org/10.1021/acs.chemmater.6b05496

[95] A.M. Ganose, C.N. Savory, D.O. Scanlon, Beyond Methylammonium Lead Iodide: Prospects for the emergent field of ns 2 containing solar absorbers, Chem. Commun. 53 (2017) 20-44. https://doi.org/10.1039/C6CC06475B

[96] T.N. Huq, L.C. Lee, L. Eyre, W. Li, R.A. Jagt, C. Kim, S. Fearn, V. Pecunia, F. Deschler, J.L.M. Driscoll, R.L.Z. Hoye, Electronic structure and optoelectronic properties of Bismuth oxyiodide robust against percent-level Iodine-, Oxygen-, and

Bismuth-related surface defects, Adv. Funct. Mater. 30 (2020) 1909983. https://doi.org/10.1002/adfm.201909983

[97] Y. Li, D.M. Lopez, V.R. Vargas, J. Zhang, K. Yang, Stability diagrams, defect tolerance, and absorption coefficients of hybrid halide semiconductors: High-throughput first-principles characterization, J. Chem. Phys. 152 (2020) 084106. https://doi.org/10.1063/1.5127929

[98] W. Gao, C. Ran, J. Xi, B. Jiao, W. Zhang, M. Wu, X. Hou, Z. Wu, High-quality $Cs_2AgBiBr_6$ double perovskite film for Lead-free inverted planar heterojunction solar cells with 2.2% efficiency, ChemPhysChem. 19 (2018) 1696-1700. https://doi.org/10.1002/cphc.201800346

[99] F. Umar, J. Zhang, Z. Jin, I. Muhammad, X. Yang, H. Deng, K. Jahangeer, Q. Hu, H. Song, J. Tang, Dimensionality controlling of $Cs_3Sb_2I_9$ for efficient all-inorganic planar thin film solar cells by HCl-assisted solution method, Adv. Opt. Mater. 7 (2019) 1801368. https://doi.org/10.1002/adom.201801368

[100] E.A. Duijnstee, J.M. Ball, V.ML. Corre, L.J.A. Koster, H.J. Snaith, J. Lim, Towards understanding space-charge limited current measurements on metal halide perovskites, ACS Energy Lett. 5 (2020) 376-384. https://doi.org/10.1021/acsenergylett.9b02720

[101] C.K. Liu, Q. Tai, N. Wang, G. Tang, H.L. Loi, F. Yan, Sn-based perovskite for highly sensitive photodetectors, Adv. Sci. 6 (2019) 1900751. https://doi.org/10.1002/advs.201900751

[102] K. Nishimura, D. Hirotani, M.A. Kamarudin, Q. Shen, T. Toyoda, S. Iikubo, T. Minemoto, K. Yoshino, S. Hayase, Relationship between lattice strain and efficiency for Sn-perovskite solar cells, ACS Appl. Mater. Interfaces. 11 (2019) 31105-31110. https://doi.org/10.1021/acsami.9b09564

[103] S.Y. Kim, Y. Yun, S. Shin, J.H. Lee, Y.W. Heo, S. Lee, Wide range tuning of band gap energy of $A_3B_2X_9$ perovskite-like halides, Scr. Mater. 166 (2019) 107-111. https://doi.org/10.1016/j.scriptamat.2019.03.009

[104] P.C. Harikesh, B. Wu, B. Ghosh, R.A. John, S. Lie, K. Thirumal, L.H. Wong, T.C. Sum, S. Mhaisalkar, N. Mathews, Doping and switchable photovoltaic effect in Lead-free perovskites enabled by metal cation transmutation, Adv. Mater. 30 (2018) 1802080. https://doi.org/10.1002/adma.201802080

[105] V. Pecunia, Y. Yuan, J. Zhao, K. Xia, Y. Wang, S. Duhm, L. Portilla, F. Li, Perovskite-inspired Lead-free Ag_2BiI_5 for self-powered NIR-blind visible light

photodetection, Nano-Micro Lett. 12 (2020) 1-12. https://doi.org/10.1007/s40820-020-0371-0

[106] F. Giustino, H.J. Snaith, Towards Lead-free perovskite solar cells, ACS Energy Lett. 1 (2016) 1233-1240. https://doi.org/10.1021/acsenergylett.6b00499

[107] M. Liu, M.B. Johnston, H.J. Snaith, Efficient planar heterojunction perovskite solar cells by vapour deposition, Nature. 501 (2013) 395-398. https://doi.org/10.1038/nature12509

[108] C. Momblona, L.G. Escrig, E. Bandiello, E.M. Hutter, M. Sessolo, K. Lederer, J.B. Nimoth, H.J. Bolink, Efficient vacuum deposited p-i-n and n-i-p perovskite solar cells employing doped charge transport layers, Energy Environ. Sci. 9 (2016) 3456-3463. https://doi.org/10.1039/C6EE02100J

[109] J. Xi, Z. Wu, B. Jiao, H. Dong, C. Ran, C. Pio, T. Lei, T.B. Song, W. Ke, T. Yokoyama, X. Hou, Multichannel interdiffusion driven FASnI3 film formation using aqueous hybrid salt/polymer solutions toward flexible Lead-free perovskite solar cells, Adv. Mater. 29 (2017) 1606964. https://doi.org/10.1002/adma.201606964

[110] D. Moghe, L. Wang, C.J. Traverse, A. Redoute, M. Sponseller, P.R. Brown, V. Bulović, R.R. Lunt, All vapor-deposited Lead-free doped CsSnBr3 planar solar cells, Nano Energy. 28 (2016) 469-474. https://doi.org/10.1016/j.nanoen.2016.09.009

[111] Jung, M. Cherl, S.R. Raga, Y. Qi, Properties and solar cell applications of Pb-free perovskite films formed by vapor deposition, RSC Adv. 6 (2016) 2819-2825. https://doi.org/10.1039/C5RA21291J

[112] N.J. Jeon, J.H. Noh, Y.C. Kim, W.S. Yang, S. Ryu, S.I. Seok, Solvent engineering for high-performance inorganic-organic hybrid perovskite solar cells, Nat. Mater. 13 (2014) 897-903. https://doi.org/10.1038/nmat4014

[113] N. Ahn, D.Y. Son, I.H. Jang, S.M. Kang, M. Choi, N.G. Park, Highly reproducible perovskite solar cells with average efficiency of 18.3% and best efficiency of 19.7% fabricated via Lewis base adduct of Lead (II) Iodide, J. Am. Chem. Soc. 137 (2015) 8696-8699. https://doi.org/10.1021/jacs.5b04930

[114] T.B. Song, Q. Chen, H. Zhou, C. Jiang, H.H. Wang, Y. Yang, Y. Liu, J. You, Y. Yang, Perovskite solar cells: Film formation and properties, J. Mater. Chem: A, 3 (2015) 9032- 9050. https://doi.org/10.1039/C4TA05246C

[115] J. Burschka, N. Pellet, S.-J. Moon, R.H. Baker, P. Gao, M.K. Nazeeruddin, M. Gr"atzel, Sequential deposition as a route to high-performance perovskite-sensitized solar cells, Nature. 499 (2013) 316-319. https://doi.org/10.1038/nature12340

[116] M. Weiss, J. Horn, C. Richter, D. Schlettwein, Preparation and characterization of Methylammonium Tin Iodide layers as photovoltaic absorbers, Phys. Status Solidi: A, 213 (2016) 975-981. https://doi.org/10.1002/pssa.201532594

[117] T. Yokoyama, D.H. Cao, C.C. Stoumpos, T.B. Song, Y. Sato, S. Aramaki, M.G. Kanatzidis, Overcoming short-circuit in Lead-free CH3NH3SnI3 perovskite solar cells via kinetically controlled gas-solid reaction film fabrication process, J. Phys. Chem. Lett. 7 (2016) 776-782. https://doi.org/10.1021/acs.jpclett.6b00118

[118] C. Wu, Q. Zhang, Y. Liu, W. Luo, X. Guo, Z. Huang, H. Ting, W. Sun, X. Zhong, S. Wei, S. Wang, The dawn of Lead-free perovskite solar cell: Highly stable double perovskite Cs2AgBiBr6 film, Adv. Sci. 5 (2018) 1700759. https://doi.org/10.1002/advs.201700759

[119] J.H. Im, H.S. Kim, N.G. Park, Morphology-photovoltaic property correlation in perovskite solar cells: One-step versus two-step deposition of CH3NH3PbI3, APL Mater. 2 (2014) 081510. https://doi.org/10.1063/1.4891275

[120] L.K. Ono, M.R. Leyden, S. Wang, Y. Qi, Organometal halide perovskite thin films and solar cells by vapor deposition, J. Mater. Chem: A, 4 (2016) 6693-6713. https://doi.org/10.1039/C5TA08963H

[121] Z. Xiao, C. Bi, Y. Shao, Q. Dong, Q. Wang, Y. Yuan, C. Wang, Y. Gao, J. Huang, Efficient, high yield perovskite photovoltaic devices grown by interdiffusion of solution-processed precursor stacking layers, Energy Environ. Sci. 7 (2014) 2619-2623. https://doi.org/10.1039/C4EE01138D

[122] D. Bi, C. Yi, J. Luo, J.D. D'ecoppet, F. Zhang, S.M. Zakeeruddin, X. Li, A. Hagfeldt, M. Gr"atzel, Polymer-templated nucleation and crystal growth of perovskite films for solar cells with efficiency greater than 21%, Nat. Energy. 1 (2016) 1-5. https://doi.org/10.1038/nenergy.2016.142

[123] Y. Deng, Q. Dong, C. Bi, Y. Yuan, J. Huang, Air-stable, efficient mixed-cation perovskite solar cells with Cu electrode by scalable fabrication of active layer, Adv. Energy Mater. 6 (2016) 1600372. https://doi.org/10.1002/aenm.201600372

[124] H. Zhang, J. Shi, X. Xu, L. Zhu, Y. Luo, D. Li, Q. Meng, Mg-doped TiO2 boosts the efficiency of planar perovskite solar cells to exceed 19%, J. Mater. Chem: A, 4 (2016) 15383-15389. https://doi.org/10.1039/C6TA06879K

[125] B. Conings, A. Babayigit, M.T. Klug, S. Bai, N. Gauquelin, N. Sakai, J.T.-W. Wang, J. Verbeeck, H.-G. Boyen, H.J. Snaith, A universal deposition protocol for planar heterojunction solar cells with high efficiency-based hybrid Lead halide

perovskite families, Adv. Mater. 28 (2016) 10701-10709.
https://doi.org/10.1002/adma.201603747

[126] F. Ye, H. Chen, F. Xie, W. Tang, M. Yin, J. He, E. Bi, Y. Wang, X. Yang, L. Han,
Soft-cover deposition of scaling-up uniform perovskite thin films for high cost-
performance solar cells, Energy Environ. Sci. 9 (2016) 2295-2301.
https://doi.org/10.1039/C6EE01411A

Perovskite based Materials for Energy Storage Devices
Materials Research Foundations 151 (2023) 155-176

Materials Research Forum LLC
https://doi.org/10.21741/9781644902738-6

Chapter 6

Technical Potential Evaluation of Inorganic Tin Perovskite Solar Cells

Lutfu Sagban Sua [1*], Figen Balo [2]

[1] Department of Management and Marketing, Southern University and A&M College

[2] Department of METE, Firat University, 23279 Elazig, Turkey

* figenbalo@gmail.com

Abstract

A component that contributes to high efficiency is the photovoltaic cell design. The perovskite photovoltaic cell is a shining star in the world of the solar panel industry. Since perovskite structure allows ions as a dopant, this path with a variety of metal cations can be pursued further. Although the subject of selecting renewable energy supplies has been explored in the literature, the challenge of selecting solar panels has only been studied in a handful of studies. The inorganic tin perovskite photovoltaic cells as the selection problem with multiple-criteria decision-making methodology have not been researched yet. In this study, the most effective solar cell among the latest inorganic tin perovskite solar cells is analyzed by AHP methodology.

Keywords

Inorganic Tin Perovskite Solar Cells, Photovoltaic, Solar Cell, Solar Energy, Renewable Energy

Contents

Technical Potential Evaluation of Inorganic Tin Perovskite Solar Cells ... 155

1. Introduction 156

2. Inorganic tin perovskite solar cells parameters used in
AHP analysis 160

3. AHP Methodology 162

4. Results and discussion 165

Conclusions 167

References 168

1. Introduction

Tin-based perovskites have recently gotten a lot of interest as a solution to the toxicity of lead halide perovskite compounds. Because tin resides in the identical major class (IV) like lead, it has been frequently reported to be used as a lead substitute in perovskite combinations for optoelectronic appliances. A thorough examination of tin-based inorganic perovskites with adjustable optical bandgap and crucial thermal stability has been initiated, and advances in photovoltaics depending on them have been noted. However, they encounter difficulties in film morphology, performance, stability, and repeatability, all of which impede commercialization. The oxidation of Sn^{2+} is an inherent and essential problem for appliances, resulting in undecided performance that falls short of their lead-based equivalents. Intensive study and the application of computational approaches, on the other hand, are undertaken to improve these qualities. Generally, the latest advancements in tin-based perovskites have created new possibilities for the sector. Solar cells utilizing a tin-sourced inorganic perovskite sheet as a light absorbent now have a maximal energy transformation performance of many more than 10%. The tin-based perovskites' reaction kinetics may be tailored for coherent generation of complete coverage, big-grained films, and pinhole-free in terms of film shape. Tin perovskite's features may be used in a range of implementations, with the inclusion of photodetectors, light-emissive diodes, sensors, and radiation detectors. According to the literature, there is a requirement to choose a solar panel of the right sort to produce power, which is dependent on the location and application needs. The selection of solar panels requires a greater number of objective and subjective elements, many of which have opposing effects. Furthermore, the precision of the selection process is determined by the solution approach utilized. Due to the aforementioned restrictions, an appropriate model for selecting a solar panel among a large number of options is required.

With the growing population and increasing industrialization, today's community's power requisitions will proceed to outgrow. Because of reverse climatic changes, with the decreasing fossil-based energy sources, the research for applicable resources of sustainable energy is continuing. Sun power is one of the most encouraging of all advanced types of sustainable power. Solar energy's source, namely sunlight, is abundant around the world, foreseeable, and many orders of magnitude greater than our required power demand. Currently, the solar market is predominated through photovoltaic cells accomplished c-Si. In itself, there has been a regulated endeavor and a more inexpensive option for Si photovoltaic cells. The performance of a crystalline Si cell in converting sunlight to electric power is roughly 20 percent. All the same, even in key diminish the light in the cost of c-Si, the installation costs and high production cause long pay-back times in many zones, descending the financial applicability of popular utilize [1]. The silicon photovoltaic cell industry's downsides, like silicon dioxide reduction, wafer fabrication, and silicon purification, have prompted a search into resolution-workable photovoltaic cell industries [2, 3]. Low-temperature synthesis, mass manufacturing, and reagent versatility are all advantages of solution processing. Perovskite photovoltaic cells, dye-sensitized, organic, and quantum dot photovoltaic cells are some of the newer solution-workable photovoltaic cells [4].

Miyasaka and co-workers published one of the earliest studies of a photovoltaic cell with perovskite absorbent in 2009, demonstrating a 3.80 percent efficiency sensitized perovskite photovoltaic cell with fluid electrolyte [5]. Park et al. [6] enhanced this efficiency to 6.50 percent [7]. The perovskite-based product, however, was melted within a few minutes during device operations because of the liquid electrolyte's corrosive nature. Within the current 2 years, this triggered a move to solid-state hole conductors and perovskite solar cell performances have risen from approximately 10.45 percent to a certified 17.90 percent [6-8].

Throughout 5 years of life, the Si photovoltaic cost has plummeted dramatically, and it is fast nearing the generating power cost of fossil-based energy sources. Photovoltaic, on the other hand, will only become the dominant energy resource if the expenses can be reduced to the point where it becomes less expensive than coal-fired energy generation. Solar cells made of inorganic–organic perovskite-based crystalline have flourished in the photovoltaic research community in the last two years, and photovoltaic cells made of these are hoped to achieve the identical productivities as c-Si at a much lower cost.

The lead-sourced perovskite photovoltaic cells' performance had achieved a plateau, with considerable energy conversion effectiveness outgrowth comparable to that of regular poly-crystalline silicon photovoltaic cells. The lead-sourced perovskites' performance in solar implementations has been exceptional, resulting in a sharp increase in perovskite cell efficiency values from 3.80 to 25.50 percent during the last decade. Despite the extraordinary accomplishments made by photovoltaic cells depending on these perovskites in the last ten years, they nevertheless have significant limitations. The stability of halide perovskites is not even close to that of Si. The photoelectron chemistry that governs their efficiency, as well as the deterioration pathways and roles of the chemicals used, are all unknown. Furthermore, the possible health and environmental threat of lead poisoning has been viewed as a disadvantage since the beginning of their use.

Although the existence of toxic water-soluble lead ingredient in perovskite solar cells is increasing relevance in academic society and a key blockage for their commercialization, lead-including perovskite photovoltaic cell industry is on the commercialization brink and has a big capacity to substitute Si solar cells. Thus, some elements like Bi (bismuth) [9], Sb (antimony) [10], Cu (copper) [11], Sn (tin) [12], and Ge (germanium) [13] are recently being examined to replace the extremely dangerous lead, because their derivatives are lesser toxicological [14]. Sn and Ge, both of group 14 metals, are the most promising substitutes for lead in perovskite production. However, it is widely known that as you move up the class fourteen elements, the 2+ oxidization status's steadiness declines, hence the main issue with using these elements is their chemical imbalance in the needed rust condition. Sn has emerged as the most viable choice for substitution among these components [15-17].

Owing to its exceptional semi-conductor features, like big carrier mobility and a plausible band gap, tin perovskite has gotten a lot of interest in recent years [18]. Tin-sourced perovskite produces, as an example, have exhibited outstanding variability in transistors, however, these products may be added to occur either purposefully or accidentally with mineral form, too. When the Sn^{2+} ions are oxidized to Sn^{4+}, the Sn^{4+} functions because a p-kind additive within produces, an operation known as "self-doping" has been established [13-15]. The photovoltaic cell with a totally without lead, tin-sourced perovskite as the absorbent sheet has never been reported before. Recently, Ogimi et al. described a blended metallict tin–lead perovskite that permitted tuning of the perovskite absorber's band-gap by changing the tin–lead proportion, displaying that tin might be a useful metal ion to use, especially for reduced band-gap photovoltaic cells [19]. But a pure $CH_3NH_3SnI_3$ perovskite doesn't display important solar capabilities and lead's minimal amount is required to stable tin in its 2+ rust condition, according to the same study.

The tin-sourced perovskite photovoltaic cells are a possible alternative to perovskite photovoltaic cells without the use of lead since they are lesser hazardous and have better photo-electric properties. The tin-sourced perovskite, like lead-sourced perovskite, has the generic structural formulation $ASnX_3$ (Here "X" is a monovalent anion, and "A" is a cation). The band-gap of $CH_3NH_3SnI_3$ (methylammonium-tin-triiodide) is from 1.2 to 1.3 eV, while the formamidinium-tin-triiodide's band-gap is 1.4 eV. Tin-sourced perovskites, which are less poisonous and have great photoelectric characteristics, clearly perform well. The tin perovskites' quick crystallization, high deficiency density, and high p-kind carrier content are among the primary challenges that must be solved to build good-performance tin perovskite photovoltaic cells. Even though Sn^{2+} and Pb^{2+} have comparable electrical configurations, Sn^{2+} has 2 energetic electrons, making tin-based perovskite low steady. Many techniques have been proposed to address these issues. Lewis soles, like C_2H_6OS, are commonly employed to limit the tin perovskite's crystallization ratio, while an oxy preservative sheet and a plethora of additions (such as hydrazine vapor, SnF_2, and fluid formic acid) have been discovered to minimize oxidization. In addition, it has been demonstrated that low-dimension structure and device engineering effectively increase the tin perovskite photovoltaic cells' efficiency. The tin-based perovskite photovoltaic cells' performance and stability have improved in recent years as a result of the aforementioned measures, indicating that tin perovskite photovoltaic cells are the best up-and-coming choice of perovskite photovoltaic cells without lead. Tin-based perovskite photovoltaic cells have now achieved a verified performance of more than 12 percent, the maximum figure for perovskite photovoltaic cells without lead. The tin-sourced perovskite photovoltaic cells are still in the testing stage, with fewer publications compared to lead-sourced perovskite photovoltaic cells. This is mostly owing to the tin's 2+ oxidization state (Sn^{2+}) instability in $CH_3NH_3SnI_3$ (methylammonium-tin-iodide), which may readily be oxidated to the more stable Sn^{4+} [20], resulting in a treatment known as self-doping [21], in which the Sn^{4+} functions as a p-dopant, lowering solar cell efficiency. For methylammonium tin iodide ($CH_3NH_3SnI_3$), the greatest recorded photovoltaic cell performance was 6.4 percent [22], 2.02 percent for $CsSnI_3$ [23, 24], 5.73 percent for $CH_3NH_3SnIBr_2$ [25], and more than 9 percent for formamidinium-tin-triiodide ($CH(NH_2)_2SnI_3$) [24, 26]. The tin-sourced perovskite photovoltaic cells' key benefits include the absence of lead and their ability to fine-adjust the active layer's band gap. There are ecological worries with employing lead-sourced perovskite photovoltaic cells in big-scale implementations [27, 28]; for example, lead is very poisonous and thus any pollution from broken photovoltaic cells might create serious environmental and health issues [29, 30]. Despite its previously reported less performance, formamidinium-tin-triiodide may hold hope since, when implemented as a thin layer, it seems to have the ability to overcome the Shockley/Queisser limit by permitting hot electron keep, thereby increasing efficiency significantly [31]. Historically, the main development of scientific studies on tin-based cells is as follows.

The organic and inorganic tin-based perovskites' synthesis was first reported in 1978 [32]. In the 1990s and beyond, scientists grew interested in their formations, physical characteristics, and conductivity. Yamada et al. used 127I-NQR and 119Sn Mössbauer spectroscopy to show how their electrical characteristics are affected by temperature [33]. Mitza et al. demonstrated that tin-based perovskites with the chemical formulation $(C_4H_9NH_3)_2(CH_3NH_3)_{n-1}Sn_nI_{3n+1}$ may be stacked and crystallized at room temperature [34]. They demonstrated the change in their behavior from semi-conductor to metallic by increasing the perovskite sheets' size, which generated a tri-dimensional structure. Mitza et al. grouped $MASnI_3$ as a less transporter p-type metal density with a density of Hall hole

(1-RHe) of (about $2 \times 10^{19}/cm^3$) and observed a decrease in resistivity with a lowering in heat [35]. Takahashi et al. demonstrated that $MASnI_3$ is a p-doped semiconductor rather than a semi-metal as previously thought. The so-called metallic transport is caused by spontaneous hole doping [36]. Among all of these characterizations, one feature that was ignored was the possibility of using these materials as solar absorbers. A hybrid Pb-Sn perovskite containing a P_3HT transportation material was described by Ogomi et al. [19]. They discovered that due to the oxidation of Sn, pure tin perovskites had no efficiency (II). The addition of Pb in various amounts, on the other hand, boosted efficiency. With $MASn_{0.5}Pb_{0.5}I_3$, an efficiency of 4.18 percent was achieved. The mesoporous $MASnI_3$ photovoltaic cell was obtained with a 6.4 percent efficiency by Noel et al. [37].

In place of oxidizing process Li bis(trifluoromethylsulfonyl)imide salt (Li/TFSI), spiro-OMeTAD was modified with H_2 bis(trifluoromethanesulfonyl)imide (H/TFSI) and epoxy encapsulation was used. The authors emphasized the need of suppressing Sn(2) oxidation to increase charge carrier span length and lower the holes' background density. Using standard spiro-OMeTAD as a hole transport material, Hao et al. produced mesoporous $CH_3NH_3SnI_3$ and $CH_3NH_3SnI_3$-xBrx perovskites solar cells [38]. Adjustment in the band gap using Br-doping to fix the open-circuit voltage resulted in a 5.7 percent efficiency. The insufficient film coverage, fill factor, and photocurrent density caused by doping of p-type through Sn(2) oxidization were the major causes of performance loss. Due to their band gap near the ideal essence for powerful optical absorption, solar implementations, outstanding charge carrier mobilities, and non-toxic material features, tin-sourced perovskites are great alternatives for producing high-efficiency photovoltaics. Nonetheless, tin halide perovskites' effectiveness is still limited by their low-cost, low defect tolerance, rapid crystallization, and oxidative instability. Despite the promising properties of tin-sourced perovskites, the rapid Sn^{+2} oxidation is a severe disadvantage. As a result, creating long-lasting devices remains challenging.

This study focused on the latest scientific findings related to inorganic tin-based perovskite photovoltaic cells. In this framework, the most efficient scientific outputs were researched for technical potential evaluation of inorganic tin-based perovskite photovoltaic cells. AHP methodology is applied to the latest developed inorganic tin perovskite solar cell types to define the most effective inorganic tin-based perovskite photovoltaic cell.

2. Inorganic tin perovskite solar cells parameters used in AHP analysis

Photovoltaic technology, although an expensive and renewable technology, is the simplest electricity generation technology in terms of design and installation. Photovoltaic cells are semiconductor materials that transform solar energy directly into electricity. The power conversion factor for any photovoltaic system is defined as efficiency. This situation causes some misconceptions. The efficiency of a photovoltaic cell can be considered as the produced electricity's rate to the whole solar radiation. According to this definition, only the electricity generated through the solar cell is taken into account, such as ambient temperature, cell temperature, and chemical composition of the photovoltaic cell. Other components and properties of photovoltaic cells are not taken into account. Photovoltaic cells are heated up during their use and loss or gain of heat according to the temperature of the environment. In other words, the heat energy released as a result of the heating of the photovoltaic cell must also be taken into account in the collection efficiency of the photovoltaic cell. In this study; production of current from photon current coming from the sun, current-voltage-power properties, filling factor, highest power point, photovoltaic cell efficiency, elements influencing power transformation performance, thermodynamic performance restriction, highest performance, quantum efficiency and factors affecting photovoltaic cell efficiency were investigated.

Among these parameters, the most important comparison characteristics were investigated. According to the findings obtained from scientific research, an analysis table was created based on the following parameters that stand out in the evaluation of photovoltaic cells. The photovoltaic cell's open-circuit voltage (Voc) gauged at the cell results when the circulation current is zero by the cell [39]. The short-circuit current (Jsc) of the photovoltaic cell is the flowing current by the cell under illumination at zero voltage. In the ideal case where the parallel resistance effects are neglected, it is equivalent to the current produced through the light and depends on the radiation intensity.

The fill factor (FF) is a variant utilized to specify the maximal throughput energy of a photovoltaic cell based on its short-circuit current and open-circuit voltage. Because the series capacitance raises, the valuation of the fill factor diminishes. The fill factor, open-circuit voltage (V_{OC}), and maximum power value (Pm) are determined by dividing them by the product of the circuit current (J_{SC}) [40]. The fill factor value is the photovoltaic cell ideality's measure. In an optimum photovoltaic cell, the fill factor is equivalent to one. Thus, it is preferred that the fill factor be close to 1 in any photovoltaic cell. For large fill factor, series resistance (Rs), diode ideality factor, temperature, and reverse saturation current density are less. The shunt resistance and energy gap restriction must be high [41]. The fill-factor value is immediately influenced by the shunt and series capacitance valuations of the diode and the cell losses. The fill factor can be raised by decreasing the series capacitance and raising the shunt capacitance. In this scenario, higher efficiency is achieved by maximizing the cell output power theoretically. The fill factor is about 80 percent for a regular silicon photovoltaic cell. It is another determining factor in the photovoltaic cell's overall behavior.

Table 1 The inorganic tin-sourced perovskite photovoltaic cells' performance with elemental substitution and diverse additives.

Device structure	Light absorber	V_{OC} (V)	J_{SC} (mA/cm^2)	PCE	FF (%)	Ref./year
FTO-c-TiO$_2$-m-TiO$_2$-{en}CsSnI$_3$-PTAA-Au	{en}CsSnI$_3$	0.97	18.75	3.79	24.8	[43]
FTO-c- TiO$_2$-CsSnI$_3$-HTM-Au	CsSnI$_3$+SnF$_2$	0.24	22.70	2.02	37	[44]
FTO-PCBM-CsSn$_{0.5}$Ge$_{0.5}$I$_3$-Spiro-OMeTAD-Au	CsSn$_{0.5}$Ge$_{0.5}$I$_3$	0.63	18.61	7.11	60.6	[45]
FTO-c-TiO$_2$-CsSnI$_3$-PTTA-Au	CsSnI$_3$+SnBr$_2$	0.44	18.5	4.30	52.9	[46]
ITO-PEDOT:PSS-CsSnI$_3$-C60-BCP-Cu	CsSnI$_3$+TSC	0.63	19.7	8.20	66.1	[47]
ITO-CsSnI$_3$-PCBM-BCP-Al	CsSnI$_3$+SnCl$_2$	0.5	9.89	3.56	68	[48]
FTO -c-TiO$_2$-m-TiO$_2$-Al$_2$O$_3$- CsSnIBr$_2$-Carbon	CsSnIBr$_2$+HPA	0.33	16.7	3.2	53	[49]
FTO -d-TiO$_2$-m-TiO$_2$-CsSnBr$_3$-spiro-MEOTAD-Au	CsSnBr$_3$+SnF$_2$	0.42	9.1	2.17	57	[50]
FTO -c-TiO$_2$-m-TiO$_2$-Al$_2$O$_3$- NiO-carbon framework	CsSnI$_3$+APZ	0.40	21.7	5.12	59	[51]
FTO -c-TiO$_2$-m-TiO$_2$-CsSnI$_3$- PTTA-Au	CsSnI$_3$+hydrazine	0.17	30.75	1.83	34.9	[52]
ITO- PEDOT: PSS-CsSnI$_3$-ICBA- BCP-Ag	CsSnI$_3$+PTM	0.64	21.81	10.1	72.1	[53]
FTO-c-TiO$_2$-m-TiO$_2$-CsSnBr$_3$-PTTA-Au	CsSnBr$_3$+hydrazine	0.37	13.96	3.04	59.4	[52]
FTO -c- TiO$_2$-m-TiO$_2$-CsSnI$_3$- HTM-Au	CsSnI$_3$+MBAA	0.45	24.85	7.5	67	[54]
FTO -c- TiO$_2$-m-TiO$_2$-CsSnI$_3$- PTTA-Au	CsSnI$_3$+SnI$_2$+hydrazine	0.38	25.71	4.81	49.1	[55]
FTO-c-TiO$_2$-m-TiO$_2$-CsSnI$_3$-PTTA-Au	CsSnI$_3$+piperazine	0.34	20.63	3.83	54.2	[56]
FTO -c- TiO$_2$-m-TiO$_2$-Al$_2$O$_3$- NiO-carbon framework	CsSnI$_3$+Co(C$_2$H$_5$)$_2$	0.36	18.24	3.0	46	[57]

The photovoltaic cell's major parameter is its capability to transform light into electricity stage. This is known as PCE (energy conversion performance) and is the rate of event light energy to produce electrical energy. To obtain the power conversion efficiency, and other beneficial metrics, I-V (current-voltage) measurements are applied [42]. The inorganic tin-sourced perovskite solar cells' performance values with elemental substitution and diverse additives are given in Table 1.

3.   AHP Methodology

A photovoltaic system's solar panel selection is a multi-criteria decision-making dilemma with both quantitative and qualitative aspects. Because it incorporates both quantitative and qualitative factors, multiple-criteria decision-making is a complicated decision-making technique. Several multiple-criteria decision-making approaches and strategies have been suggested in recent years to manipulate energy planning concerns. Sasikumar et al. provided an arithmetical strategy for choosing steel facility suppliers utilizing, a decision support system, and a fuzzy AHP was created to evaluate supplier performance using the integrated approach [58]. To design energy management strategies, Cai et al. suggested the use of arithmetical modeling depending on an interactive decision support system [59]. Mardani et al. undertook a strategic assessment of the use of decision-making approaches for power management challenges [60]. Charabi et al. utilized a synthesis of fuzzy techniques and GIS to determine the feasibility of establishing solar photovoltaic power facilities in Oman [61]. Sapuan et al. [62] used a knowledge engineering environment to choose composite materials for application in the vehicle sector. Kahraman et al. suggested a fuzzy AHP-based selection model for renewable energy sources [63]. For energy concerns, Kowalski et al. recommended a participant multiple-criteria decision-making approach [64]. Ghazanfar Khan et al. devised a method for locating potential solar photovoltaic plant sites [65]. For supplier selection, Wang and Ho et al. used fuzzy hierarchic TOPSIS and TOPSIS [66, 67]. Fuzzy TOPSIS was used by Anjali et al. to pick up the best alternative transportation systems [68]. Patlitzianas et al. proposed a DSS for defining the environment of contemporary energy businesses [69]. Vetrivelsezhian et al. suggested novel modeling depending on AHP and TOPSIS to analyze the performance of transportation corporations [70]. Cavallaro devised the Fuzzy-TOPSIS methodology for assessing photovoltaic facility heat storage [71]. By merging VIKOR-AHP approaches, Kaya et al. created multi-criteria modeling to choose the best sustainable power option [72]. Wang et al. [73] published a review on sustainable power decision support systems. Tavana et al. improved one decision-support system for solar farm site planning [74]. By combining fuzzy AHP and TOPSIS models, Kengpol et al. presented a decision-support system to minimize flooding on photovoltaic energy facility site choice [75]. In Malaysia, Amın et al. documented solar cell efficiency [76]. For measuring photovoltaic network potency, Ramachandra et al. created a decision support system with user-friendly properties [77]. Beltran chose photovoltaic projects using an analytic network technique

[78]. An ANP model was developed by O nut et al. to undertake a multiple-criteria evaluation to pick optimal energy resources [79].

The analytic hierarchy process is a tool for MCDM. The pairwise crosschecks are made utilizing an Eigen-value technique. It contains an operation for adjusting the numerical measure for both qualitative and quantitative efficiency assessments as well. The measure varies from one-nine for "minimum value than", to one for "equivalent", and nine for "completely more value than", spanning the complete benchmark spectrum. Saaty (1980) developed this approach at the University of Pittsburgh. The analytic hierarchy process provides a multiple-criteria context for assessing options. That is, it is theoretically sound and relatively simple. The analytic hierarchy process enables decision-makers to cope with complicated troubles utilizing a fundamental organizational formation in addition to examining both qualitative and quantitative information in a methodical route under many competing criteria [80].

Under the four steps of decomposition, pairwise comparisons, priority vector creation, and synthesis [81, 82], the AHP finally achieves an ideal alternative.

- Identify the issue.
- Extend the scope of the problem's goals and outcomes.
- Determine the factors that impact the goals.
- Break the multi-criteria trouble down into an organizational formation with several tiny constituent pieces.
- Using Saaty's nine-point scale, a sequence of pairwise crosschecks are made amongst the factors at the identical ratio at the future better ratio.
- Create the precedence carrier at each level according to each of the elements using the eigenvector approach. Synthesizing all of the priority vectors yields a final priority vector of choices.

According to the literature analysis, the following factors make the decision-making process for solar panel selection tough and complicated:

- A variety of qualitative and quantitative criteria.
- Criteria clashes.
- A wide range of options.
- The buying process is subjected to both external and internal limitations.

Allowing managers to be flexible in varied situations because material managers must manage supply functions in the face of increased environmental unpredictability as well as fast technical, economic, and political changes on a global scale. AHP can assist managers in the selection of solar panels in the below pathways:

- Examining the influence of various resources on a company's various objectives.

- Simplifying the inputs' interactional flow from overall ratios of management to assess the resources from a planned standpoint.

- Simplifying the information' interactional flow from multi-hierarchical ratios, as photovoltaic panel choice is usually decided after numerous error and trial cycles. By modeling multiple outcomes, the AHP modeling can aid in finding a satisfactory conclusion operation.

4. Results and discussion

The decision matrix provided in Table 2 is created through the pair-wise comparison of the selected criteria. The table presents the relative priority values of the utilized criteria in the study.

Table 2. Decision Matrix

Matrix		PCE (%)	FF (%)	J_sc (mA cm−2)	V_oc (V)	Relative Priorities
		1	2	3	4	
PCE (%)	1	1	5	2	2	**48,88%**
FF (%)	2	1/5	1	3	3	**24,69%**
J_sc (mA cm−2)	3	1/2	1/3	1	2	**15,25%**
V_oc (V)	4	1/2	1/3	1/2	1	**11,17%**

Based on the final relative priority values illustrated in Figure 1, PCE (%) is concluded to have the largest impact on the selection of the inorganic tin perovskite solar cells.

For this study, 16 inorganic tin-sourced PSC alternatives have been determined. The top portion of the table provides the alternatives' normalized values, associated priority values for every attribute, and finally the total scores. Original performance values in Table 3 are divided by the sum of each criterion value to calculate the normalized values. Finally, multiplying each performance value with the associated criteria weights result in the priority values provided in the bottom portion of the table. The last row indicates the total score for each PSC alternative.

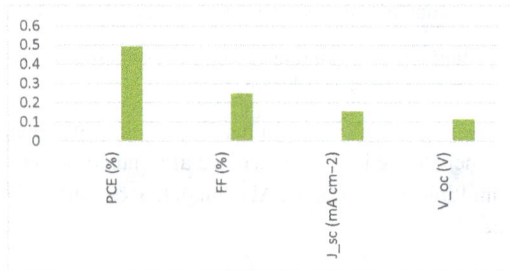

Figure 1. Criteria Priorities

Table 3. Normalized Priorities

	P1	P2	P3	P4	P5	P6	P7	P8	P9	P10	P11	P12	P13	P14	P15	P16
Normalized																
PCE (%)	0,052	0,027	0,097	0,058	0,111	0,048	0,043	0,029	0,070	0,025	0,137	0,041	0,102	0,065	0,052	0,041
FF (%)	0,029	0,043	0,070	0,061	0,077	0,079	0,062	0,066	0,069	0,041	0,084	0,069	0,078	0,057	0,063	0,053
J_sc (mA cm−2)	0,060	0,073	0,060	0,059	0,063	0,032	0,054	0,029	0,070	0,099	0,070	0,045	0,080	0,083	0,066	0,059
V_oc (V)	0,133	0,033	0,087	0,061	0,087	0,069	0,045	0,058	0,055	0,023	0,088	0,051	0,062	0,052	0,047	0,050
Priorities																
PCE (%)	0,025	0,013	0,047	0,029	0,054	0,024	0,021	0,014	0,034	0,012	0,067	0,020	0,050	0,032	0,025	0,020
FF (%)	0,007	0,011	0,017	0,015	0,019	0,019	0,015	0,016	0,017	0,010	0,021	0,017	0,019	0,014	0,016	0,013
J_sc (mA cm−2)	-0,009	-0,011	-0,009	-0,009	-0,010	-0,005	-0,008	-0,004	-0,011	-0,015	-0,011	-0,007	-0,012	-0,013	-0,010	-0,009
V_oc (V)	0,015	0,004	0,010	0,007	0,010	0,008	0,005	0,006	0,006	0,003	0,010	0,006	0,007	0,006	0,005	0,006
Total Score	0,038	0,017	0,065	0,041	0,073	0,046	0,033	0,033	0,046	0,010	**0,087**	0,036	0,064	0,039	0,036	0,030

The results in Table indicate that "ITO / PEDOT: PSS / CsSnI3 / ICBA / BCP /Ag" (P11) is the alternative with the top score (0,087) as it is illustrated in Figure 2 below.

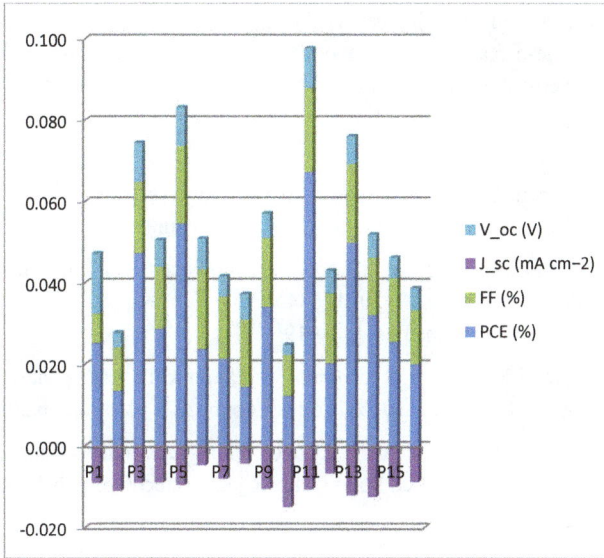

Figure 2. Alternative Scores

Conclusions

Photovoltaic technology is still plagued by the risk of lead poisoning, which puts doubt on its future industrialization. Solar cells made of inorganic lead halide perovskite have made significant progress in converting solar light into energy at a low cost. Tin perovskite solar cells, which are environmentally friendly, have currently emerged as a possible contender for perovskite photovoltaic cells without lead. With progressed research, tin perovskite photovoltaic cells have sparked a lot of interest as a potential perovskite photovoltaic cell contender without lead. Due to the oxidization of Sn^{2+} to Sn^{4+} and fewer flaws occurrence power in the course of the tin-based perovskites' crystallization, it is difficult to generate effective and stable tin-sourced perovskite photovoltaic cells. Endeavors to improve stability and efficiency have been attempted, including the use of a two/three-dimensional hierarchical formation, reducing agents, and antioxidants. Furthermore, it has been demonstrated that mixed-cation engineering may stabilize the perovskite phase. As the solar cells sent to certified associations are needed to access a stable result and go through ingeminated voltage–current analysis till the results gathered under operating circumstances are steady in time, the tin perovskite solar cell's performance's authoritative stabilized certification is still missing. Because there are no stabilized verified findings, the subject of variances in the tin perovskite photovoltaic cells' evaluation remains unanswered. For these reasons, it is important to overcome the stability and efficiency limitations of tin perovskite photovoltaic cells for future applications.

References

[1] C. Steven, M. Arun, Opportunities and challenges for a sustainable energy future, Nature. 488 (2012) 294-303. https://doi.org/10.1038/nature11475

[2] G. Bye, B. Ceccaroli, Solar grade silicon: Technology status and industrial trends, Sol. Energ. Mater. Sol. Cell. 130 (2014) 63446. https://doi.org/10.1016/j.solmat.2014.06.019

[3] Z. Dong, Y. Lin, Ultra-thin wafer technology and applications: A review. Mater. Sci. Semicond. Process. 105 (2020) 104681. https://doi.org/10.1016/j.mssp.2019.104681

[4] M.A. Mingsukang, M.H. Buraidah, A.K. Arof, Third-generation-sensitized solar cells, in: N. Das (Eds.), Nanostructured Solar Cells. Rijeka: IntechOpen, 2017, 65290. https://doi.org/10.5772/65290

[5] A. Kojima, K. Teshima, Y. Shirai, T. Miyasaka, Organometal halide perovskites as visible-light sensitizers for photovoltaic cells, J. Am. Chem. Soc. 131 (2009) 6050-6051. https://doi.org/10.1021/ja809598r

[6] J.H. Im, C.R. Lee, J.W. Lee, S.W. Park, N.G. Park, 6.5% efficient perovskite quantum-dot-sensitized solar cell, Nanoscale. 3 (2011) 4088-4093. https://doi.org/10.1039/c1nr10867k

[7] J.H. Noh, S.H. Im, J.H. Heo, T.N. Mandal, S.I. Seok, Chemical management for colorful, efficient, and stable inorganic-organic hybrid nanostructured solar cells, Nano Lett. 13 (2013) 1764-1769. https://doi.org/10.1021/nl400349b

[8] NREL, (NREL). http://www.nrel.gov/ncpv/images/efficiency_chart.jpg.

[9] F. Bai, Y. Hu, Y. Hu, T. Qiu, X. Miao, S. Zhang, Lead-free, air-stable ultrathin Cs3Bi2I9 perovskite nanosheets for solar cells, Sol. Energy Mater. Sol. Cells. 184 (2018) 15-21. https://doi.org/10.1016/j.solmat.2018.04.032

[10] B. Saparov, F. Hong, J.P. Sun, H.S. Duan, W. Meng, S. Cameron, I.G. Hill, Y. Yan, D.B. Mitzi, Thin-film preparation and characterization of Cs3Sb2I9: A lead-free layered perovskite semiconductor, Chem. Mater. 27 (2015) 5622-5632. https://doi.org/10.1021/acs.chemmater.5b01989

[11] D. Cortecchia, H.A. Dewi, J. Yin, A. Bruno, S. Chen, T. Baikie, P.P. Boix, M. Gratzel, S. Mhaisalkar, C. Soci, N. Mathews, Lead-free MA2CuCl(x)Br(4-x) hybrid

perovskites, Inorg. Chem. 55 (2016) 1044-1052.
https://doi.org/10.1021/acs.inorgchem.5b01896

[12] N.K. Noel, S.D. Stranks, A. Abate, C. Wehrenfennig, S. Guarnera, A.A. Haghighirad, A. Sadhanala, G.E. Eperon, S.K. Pathak, M.B. Johnston, A. Petrozza, L.M. Herz, H.J. Snaith, Lead-free organic-inorganic tin halide perovskites for photovoltaic applications, Energy Environ. Sci. 7 (2014) 3061-3068. https://doi.org/10.1039/C4EE01076K

[13] T. Krishnamoorthy, H. Ding, C. Yan, W.L. Leong, T. Baikie, Z. Zhang, M. Sherburne, S. Li, M. Asta, N. Mathews, S.G. Mhaisalkar, Lead-free germanium iodide perovskite materials for photovoltaic applications, J. Mater. Chem: A, 3 (2015) 23829-23832. https://doi.org/10.1039/C5TA05741H

[14] J. Li, H.L. Cao, W.B. Jiao, Q. Wang, M. Wei, I. Cantone, J. Lu, A. Abate, Biological impact of lead from halide perovskites reveals the risk of introducing a safe threshold, Nat. Commun. 11 (2020) 310. https://doi.org/10.1038/s41467-019-13910-y

[15] D.B. Mitzi, C.A. Feild, Z. Schlesinger, R.B. Laibowitz, Transport, optical, and magnetic properties of the conducting halide perovskite CH3NH3SnI3, J. Solid State Chem. 114 (1995) 159-163. https://doi.org/10.1006/jssc.1995.1023

[16] Y. Takahashi, H. Hasegawa, Y. Takahashi, T. Inabe, Hall mobility in tin iodide perovskite CH3NH3SnI3: Evidence for a doped semiconductor, J. Solid State Chem. 205 (2013) 39-43. https://doi.org/10.1016/j.jssc.2013.07.008

[17] I. Chung, J.H. Song, J. Im, J. Androulakis, C.D. Malliakas, H. Li, A.J. Freeman, J.T. Kenney, M.G. Kanatzidis, CsSnI3: Semiconductor or metal? High electrical conductivity and strong near-infrared photoluminescence from a single material. High hole mobility and phase-transitions, J. Am. Chem. Soc. 134 (2012) 8579-8587. https://doi.org/10.1021/ja301539s

[18] W. Ke, M.G. Kanatzidis, Prospects for low-toxicity Lead-free perovskite solar cells, Nat. Commun. 10 (2019) 965. https://doi.org/10.1038/s41467-019-08918-3

[19] Y. Ogomi, A. Morita, S. Tsukamoto, T. Saitho, N. Fujikawa, Q. Shen, CH3NH3SnxPb(1-x)I3 perovskite solar cells covering up to 1060 nm, J. Phys. Chem. Lett. 5 (2014)1004-1011. https://doi.org/10.1021/jz5002117

[20] S.J. Lee, S.S. Shin, Y.C. Kim, D. Kim, T.K. Ahn, J.H. Noh, J. Seo, S. Seok, Fabrication of efficient Formamidinium Tin Iodide perovskite solar cells through SnF2-Pyrazine complex, J. Am. Chem. Soc. 14 (2016) 3974-3977. https://doi.org/10.1021/jacs.6b00142

[21] Y. Takahashi, R. Obara, Z.Z. Lin, Y. Takahashi, T. Naito, T. Inabe, S. Ishibashi, K. Terakura, Charge-transport in Tin-Iodide perovskite CH3NH3SnI3: Origin of high conductivity, Dalton Trans. 40 (2011) 5563-5568. https://doi.org/10.1039/c0dt01601b

[22] N.K. Noel, S.D. Stranks, A. Abate, C. Wehrenfennig, S. Guarnera, A.-A. Haghighirad, A. Sadhanala, G.E. Eperon, S.K. Pathak, M.B. Johnston, A. Petrozza, L.M. Herz, H.J.Snaith, Lead-free organic-inorganic Tin halide perovskites for photovoltaic applications, Energy Environ. Sci. 7 (2014) 3061-3068. https://doi.org/10.1039/C4EE01076K

[23] M.H. Kumar, S. Dharani, W.L. Leong, P.P. Boix, R.R. Prabhakar, T. Baikie, C. Shi, H. Ding, R. Ramesh, M. Asta, M. Graetzel, S.G. Mhaisalkar, N. Mathews, Lead-free halide perovskite solar cells with high photocurrents realized through vacancy modulation, Adv. Mater. 26 (2014) 7122-7127. https://doi.org/10.1002/adma.201401991

[24] S. Shao, J. Liu, G. Portale, H.H. Fang, G.R. Blake, G.H.T. Brink, L.J.A. Koster, M.A. Loi, Highly reproducible Sn-based hybrid perovskite solar cells with 9% efficiency, Adv. Eng. Mater. 8 (2018) 1702019. https://doi.org/10.1002/aenm.201702019

[25] F. Hao, C.C. Stoumpos, D.H. Cao, R.P.H. Chang, M.G. Kanatzidis, Lead-free solid-state organic-inorganic halide perovskite solar cells, Nat. Photon. 8 (2014) 489-494. https://doi.org/10.1038/nphoton.2014.82

[26] E. Jokar, C.H. Chien, C.M. Tsai, A. Fathi, E.W.G. Diau, Robust tin-based perovskite solar cells with hybrid organic cations to attain efficiency approaching 10%, Adv. Mat. 31 (2018) 1804835. https://doi.org/10.1002/adma.201804835

[27] N. Espinosa, L.S.-Lujan, A. Urbina, F.C. Krebs, Solution and vapor deposited lead perovskite solar cells: Ecotoxicity from a life cycle assessment perspective, Sol. Energy Mater. Sol. Cells. 137 (2015) 303-310. https://doi.org/10.1016/j.solmat.2015.02.013

[28] J. Zhang, X. Gao, Y. Deng, B. Li, C. Yuan, Life cycle assessment of titania perovskite solar cell technology for sustainable design and manufacturing, ChemSusChem. 8 (2015) 3882-3891. https://doi.org/10.1002/cssc.201500848

[29] I.R. Benmessaoud, A.-L.M. Mellier, E. Horvath, B. Maco, M. Spina, H.A. Lashuel, L. Forro, Health hazards of Methylammonium Lead Iodide based perovskites: Cytotoxicity studies, Toxicol. Res. 5 (2016) 407-419. https://doi.org/10.1039/C5TX00303B

[30] A. Babayigit, D.D. Thanh, A. Ethirajan, J. Manca, M. Muller, H.-G. Boyen, B. Conings, Assessing the toxicity of Pb-and Sn-based perovskite solar cells in model organisms Danio rerio, Sci. Rep. 6 (2016) 18721. https://doi.org/10.1038/srep18721

[31] H.H. Fang, A. Sampson, S. Shao, J. Even, M.A. Loi, Long-lived hot-carrier light emission and large blue shift in Formamidinium Tin Triiodide perovskites, Nature Commu. 9 (2018) 243. https://doi.org/10.1038/s41467-017-02684-w

[32] D. Weber, CH3NH3SnBrxI3-x (x=0-3), ein Sn(II)-system mit kubischer Perowskitstruktur/CH3NH3SnBrxI3-x (x=0-3), a Sn(II)-system with cubic perovskite structure, Z Naturforsch: B, 33 (1978) 862-865. https://doi.org/10.1515/znb-1978-0809

[33] K. Yamada, T. Matsui, T. Tsuritani, T. Okuda, S. Ichiba, 127I-NQR, 119 Sn Mössbauer effect, and electrical conductivity of MSnI3 (M = K, NH4, Rb, Cs, and CH3NH3), Z Naturforsch: A, 45 (1990) 307-312. https://doi.org/10.1515/zna-1990-3-416

[34] D.B. Mitzi, C.A. Feild, W.T.A. Harrison, A.M. Guloy, Conducting tin halides with a layered organic-based perovskite structure, Nature. 369 (1994) 467-469. https://doi.org/10.1038/369467a0

[35] D.B. Mitzi, C.A. Feild, Z. Schlesinger, R.B. Laibowitz, Transport, optical, and magnetic properties of the conducting halide perovskite CH3NH3SnI3, J. Solid State Chem. 114 (1995) 159-163. https://doi.org/10.1006/jssc.1995.1023

[36] Y. Takahashi, R. Obara, Z.Z. Lin, Y. Takahashi, T. Naito, T. Inabe, Charge-transport in Tin-Iodide perovskite CH3NH3SnI3: Origin of high conductivity, Dalton Trans. 40 (2011) 5563-5568. https://doi.org/10.1039/c0dt01601b

[37] N.K. Noel, S.D. Stranks, A. Abate, C. Wehrenfennig, S. Guarnera, A.A. Haghighirad, Lead-free organic-inorganic Tin halide perovskites for photovoltaic applications, Energy Environ Sci. 7 (2014) 3061. https://doi.org/10.1039/C4EE01076K

[38] F. Hao, C.C. Stoumpos, D.H. Cao, R.P.H. Chang, M.G. Kanatzidis, Lead-free solid-state organic-inorganic halide perovskite solar cells, Nat. Photonics. 8 (2014) 489-94. https://doi.org/10.1038/nphoton.2014.82

[39] N.D. Arora, J.R. Hauser, Temperature dependence of Silicon solar cell characteristics, Sol. Energy Mater. 6 (1982) 151-158. https://doi.org/10.1016/0165-1633(82)90016-8

[40] M. Chegaar, A. Hamzaoui, P. Petit, M. Aillerie, A. Herguth, Effect of illumination intensity on solar cells parameters, Energy Procedia. 36 (2013) 722-729. https://doi.org/10.1016/j.egypro.2013.07.084

[41] S. Dubey, J.N. Sarvaiya, B. Seshadri, Temperature dependent photovoltaic (PV) efficiency and its effect on PV production in the world: A review, Energy Procedia. 33 (2013) 311-321. https://doi.org/10.1016/j.egypro.2013.05.072

[42] N.D. Arora, J.R. Hauser, Temperature dependence of Silicon solar cell characteristics, Sol. Energy Mater. 6 (1982) 151-158. https://doi.org/10.1016/0165-1633(82)90016-8

[43] W. Ke, C. Stoumpos, I. Spanopoulos, L. Mao, M. Chen, M. Wasielewski, M. Kanatzidis, Efficient Lead-free solar cells based on hollow {en}MASnI3 perovskites, J. Am. Chem. Soc. 139 (2017) 14800-14806. https://doi.org/10.1021/jacs.7b09018

[44] M. Kumar, S. Dharani, W. Leong, P. Boix, R. Prabhakar, T. Baikie, C. Shi, H. Ding, R. Ramesh, M. Asta, M. Graetzel, S. Mhaisalkar, N. Mathews, Lead- free halide perovskite solar cells with high photocurrents realized through vacancy modulation, Adv. Mater. 26 (2014) 7122-7127. https://doi.org/10.1002/adma.201401991

[45] M. Chen, M. Ju, H. Garces, A. Carl, L. Ono, Z. Hawash, Y. Zhang, T. Shen, Y. Qi, R. Grimm, D. Pacifici, X. Zeng, Y. Zhou, N. Padture, Highly stable and efficient all-inorganic Lead-free perovskite solar cells with native-oxide passivation, Nat. Commun. 10 (2019) 16 https://doi.org/10.1038/s41467-018-07951-y

[46] J. Heo, J. Kim, H. Kim, S. Moon, S. Im, K. Hong, Roles of SnX2 (X F, Cl, Br) additives in Tin-based halide perovskites toward highly efficient and stable Lead-free perovskite solar cells, J. Phys. Chem. Lett. 9 (2018) 6024-6031. https://doi.org/10.1021/acs.jpclett.8b02555

[47] H. Ban, T. Zhang, X. Gong, Q. Sun, X. Zhang, N. Pootrakulchote, Y. Shen, M. Wang, Fully inorganic CsSnI3 mesoporous perovskite solar cells with high efficiency and stability via Coadditive engineering, Solar RLL. 5 (2021) 2100069. https://doi.org/10.1002/solr.202100069

[48] K. Marshall, M. Walker, R. Walton, R. Hatton, Enhanced stability and efficiency in hole-transport-layer-free CsSnI3 perovskite photovoltaics, Nat. Energy. 1 (2016) 16178. https://doi.org/10.1038/nenergy.2016.178

[49] W. Li, J. Li, J. Li, J. Fan, Y. Mai, L. Wang, Addictive-assisted construction of all-inorganic CsSnIBr2 mesoscopic perovskite solar cells with superior thermal stability

up to 473 K, J. Mater. Chem. 4 (2016) 17104-17110.
https://doi.org/10.1039/C6TA08332C

[50] S. Gupta, T. Bendikov, G. Hodes, D. Cahen, CsSnBr3, A lead-free halide perovskite
for long-term solar cell application: Insights on SnF2 addition, ACS Energy Lett. 1
(2016) 1028-1033. https://doi.org/10.1021/acsenergylett.6b00402

[51] B. Li, H. Di, B. Chang, R. Yin, L. Fu, Y. Zhang, L. Yin, Efficient passivation
strategy on Sn related defects for high performance all-inorganic CsSnI3 perovskite
solar cells, Adv. Funct. Mater. 31 (2021) 2007447.
https://doi.org/10.1002/adfm.202007447

[52] T. Song, T. Yokoyama, C. Stoumpos, J. Logsdon, D. Cao, M. Wasielewski, S.
Aramaki, M. Kanatzidis, Importance of reducing vapor atmosphere in the fabrication
of Tin-based perovskite solar cells, J. Am. Chem. Soc. 139 (2017) 836-842.
https://doi.org/10.1021/jacs.6b10734

[53] T. Ye, X. Wang, K. Wang, S. Ma, D. Yang, Y. Hou, J. Yoon, K. Wang, S. Priya,
Localized electron density engineering for stabilized B-γ CsSnI3-based perovskite
solar cells with efficiencies >10%, ACS Energy Lett. 6 (2021) 1480-1489.
https://doi.org/10.1021/acsenergylett.1c00342

[54] T. Ye, K. Wang, Y. Hou, D. Yang, N. Smith, B. Magill, J. Yoon, R. Mudiyanselage,
G. Khodaparast, K. Wang, S. Priya, Ambient-air-stable lead-free CsSnI3 solar cells
with greater than 7.5% efficiency, J. Am. Chem. Soc. 143 (2021) 4319-4328.
https://doi.org/10.1021/jacs.0c13069

[55] T. Song, T. Yokoyama, S. Aramaki, M. Kanatzidis, Performance enhancement of
Lead-free Tin-based perovskite solar cells with reducing atmosphere-assisted
dispersible additive, ACS Energy Lett. 2 (2017) 897-903.
https://doi.org/10.1021/acsenergylett.7b00171

[56] T. Song, T. Yokoyama, J. Logsdon, M. Wasielewski, S. Aramaki, M. Kanatzidis,
Piperazine suppresses self-doping in CsSnI3 perovskite solar cells, ACS Appl. Energy
Mater. 1 (2018) 4221-4226. https://doi.org/10.1021/acsaem.8b00866

[57] T. Zhang, H. Li, H. Ban, Q. Sun, Y. Shen, M. Wang, Efficient CsSnI3-based
inorganic perovskite solar cells based on a mesoscopic metal oxide frame-work via
incorporating a donor element, J. Mater. Chem. 8 (2020) 4118-4124.
https://doi.org/10.1039/C9TA11794F

[58] G. Sasikumar, S. Velappan, P. Manimaran, An integrated model for supplier selection using fuzzy analytical hierarchy process: A steel plant case study, Int. J. Procure. Manag. 3 (2010) 292-315. https://doi.org/10.1504/IJPM.2010.033447

[59] Y.P. Cai, G.H. Huang, Q.G. Lin, X.H. Nie, Q. Tan, An optimization-model-based interactive decision support system for regional energy management systems planning under uncertainty, Expert Syst. Appl. 36 (2009) 3470-3482. https://doi.org/10.1016/j.eswa.2008.02.036

[60] A. Mardani, A. Jusoh, E.K. Zavadskas, Z. Khalifah, Application of multiple criteria decision making techniques in tourism and hospitality industry: A systematic review, Transform. Bus. Econ. 15 (2016) 37.

[61] Y. Charabi, A. Gastli, GIS assessment of large CSP plant in Duqm, Oman, Renew. Sustain. Energy Rev. 14 (2010) 835-841. https://doi.org/10.1016/j.rser.2009.08.019

[62] S.M. Sapuan, H.S. Abdalla, A prototype knowledge-based system for the material selection of polymeric-based composites for automotive components, Compos. Part A: Appl. Sci. Manuf. 29 (1998) 731-742. https://doi.org/10.1016/S1359-835X(98)00049-9

[63] C. Kahraman, I. Kaya, S. Cebi, A comparative analysis for multiattribute selection among renewable energy alternatives using fuzzy axiomatic design and fuzzy analytic hierarchy process, Energy. 34 (2019) 1603-1616. https://doi.org/10.1016/j.energy.2009.07.008

[64] K. Kowalski, S. Stagl, R. Madlener, I. Omann, Sustainable energy futures: Methodological challenges in combining scenarios and participatory multi-criteria analysis, Eur. J. Oper. Res. 197 (2017) 1063-1074. https://doi.org/10.1016/j.ejor.2007.12.049

[65] G. Khan, S. Rathi, Optimal site selection for solar PV power plant in an Indian state using geographical information system (GIS), Int. J. Emerg. Eng. Res. Technol. 2 (2014) 260-266.

[66] W.J. Wang, C.H. Cheng, H.K. Cheng, Fuzzy hierarchical TOPSIS for supplier selection, Appl. Soft Comput. 9 (2009) 377-386. https://doi.org/10.1016/j.asoc.2008.04.014

[67] W. Ho, X. Xu, P.K. Dey, Multi-criteria decision making approaches for supplier evaluation and selection: A literature review, Eur. J. Oper. Res. 202 (2010) 16-24. https://doi.org/10.1016/j.ejor.2009.05.009

[68] A. Awasthi, S.S. Chauhan, H. Omrani, Application of fuzzy TOPSIS in evaluating sustainable transportation systems, Int. J. Expert Syst. Appl. 38 (2011) 12270-12280. https://doi.org/10.1016/j.eswa.2011.04.005

[69] K.D. Patlitzianas, A. Pappa, J. Psarras, An information decision support system towards the formulation of modern energy companies, Environ. Renew. Sustain. Energy Rev. 12 (2008) 790-806. https://doi.org/10.1016/j.rser.2006.10.014

[70] M. Vetrivelsezhian, C. Muralidharan, T. Nambirajan, S.G. Deshmukh, Performance measurement in a public sector passenger bus transport company using fuzzy TOPSIS, fuzzy AHP and ANOVA: A case study, Int. J. Eng. Sci. Technol. 3 (2011) 1046-1059.

[71] F. Cavallaro, Fuzzy TOPSIS approach for assessing thermal-energy storage in Concentrated Solar Power (CSP) systems, Appl. Energy. 87 (2010) 496-503. https://doi.org/10.1016/j.apenergy.2009.07.009

[72] T. Kaya, C. Kahraman, Multicriteria renewable energy planning using an integrated fuzzy VIKOR & AHP methodology: The case of Istanbul, Energy. 35 (2010) 2517-2527. https://doi.org/10.1016/j.energy.2010.02.051

[73] J.J. Wang, Y.J. Jing, C.F. Zhang, J.H. Zhao, Review on multicriteria decision analysis aid in sustainable energy decisionmaking, Renew. Sustain. Energy Rev. 13 (2009) 2263-2278. https://doi.org/10.1016/j.rser.2009.06.021

[74] M. Tavana, F.J.S. Arteaga, S. Mohammadi, M. Alimohammadi, A fuzzy multi-criteria spatial decision support system for solar farm location planning, Energy Strategy Rev. 18 (2017) 93-105. https://doi.org/10.1016/j.esr.2017.09.003

[75] A. Kengpol, P. Rontlaong, M. Tuominen, A decision support system for selection of solar power plant locations by applying fuzzy AHP and TOPSIS: An empirical study, J. Softw. Eng. Appl. 6 (2013) 470-481. https://doi.org/10.4236/jsea.2013.69057

[76] N. Amin, C.W. Lung, K. Sopian, A practical field study of various solar cells on their performance in Malaysia, Renew. Energy J. 34 (2009) 1939-1946. https://doi.org/10.1016/j.renene.2008.12.005

[77] T.V. Ramachandra, R.K. Jha, S.V. Krishna, B.V. Shruthi, Solar energy decision support system, Int. J. Sustain. Energy. 24 (2005) 207-224. https://doi.org/10.1080/14786450500292105

[78] P. Beltran, An AHP (analytic hierarchy process)/ANP (analytic network process)-based multi-criteria decision approach for the selection of solar-thermal power plant investment project, Energy 17 (2013) 645-658.

[79] S.O. Nut, U.R. Tuzkaya, N. Saadet, Multiple criteria evaluation of current energy resources for the Turkish manufacturing industry, Energy Convers. Manag. 49 (2008) 1480-1492. https://doi.org/10.1016/j.enconman.2007.12.026

[80] W.B. Lee, H. Lau, Z.Z. Liu, S. Tam, Fuzzy analytic hierarchy process approach in modular product design, Expert Systems. 18 (2001) 32-42. https://doi.org/10.1111/1468-0394.00153

[81] T.L. Saaty, The Analytic Hierarchy Process, New York: McGraw-Hill, 1980. https://doi.org/10.21236/ADA214804

[82] T.V. Ramachandra, R.K. Jha, S.V. Krishna, B.V. Shruthi, Solar energy decision support system, Int. J. Sustain. Energy. 24 (2005) 207-224. https://doi.org/10.1080/14786450500292105

Keyword Index

Capacitors   67
Cost Analysis   89

Dielectrics   67

Ferroelectricity   67
Fuel Cells   67

Germanium Based   111

Halide Double Perovskites   111
High Storage Density   67
Hole Transport Materials   1

Inorganic Tin Perovskite Solar Cells 155

Lead Toxicity   89
Lead-Free   111
Lectron Transport Layer   1

One-Step Deposition Perovskites   1
Organometallic Halide-Based Perovskite
    Solar Cells   33
Organometallic Materials   1

Passivation Methods   33
Perovskite Solar Cells (PSCs)   67
Perovskite Solar Cells   33, 89
Photovoltaic   111, 155
Power Conversion Efficiency   33

Recycling Process   89
Renewable Energy   155

Solar Cell   111, 155
Solar Energy   33, 155
Spin Coating   1

Tin-Based   111
Transport Properties   67

About the Editors

Dr. Inamuddin is working as an Assistant Professor at the Department of Applied Chemistry, Aligarh Muslim University, Aligarh, India. He obtained a Master of Science degree in Organic Chemistry from Chaudhary Charan Singh (CCS) University, Meerut, India, in 2002. He received his Master of Philosophy and Doctor of Philosophy degrees in Applied Chemistry from Aligarh Muslim University (AMU), India, in 2004 and 2007, respectively. He has extensive research experience in multidisciplinary fields of Analytical Chemistry, Materials Chemistry, and Electrochemistry and, more specifically, Renewable Energy and Environment. He has worked on different research projects as a project fellow and senior research fellow funded by the University Grants Commission (UGC), Government of India, and the Council of Scientific and Industrial Research (CSIR), Government of India. He has received the Fast Track Young Scientist Award from the Department of Science and Technology, India, to work in the area of bending actuators and artificial muscles. He has also received the Sir Syed Young Researcher of the Year Award 2020 from Aligarh Muslim University. He has completed four major research projects sanctioned by the University Grant Commission, Department of Science and Technology, Council of Scientific and Industrial Research, and Council of Science and Technology, India. He has published 210 research articles in international journals of repute and nineteen book chapters in knowledge-based book editions published by renowned international publishers. He has published 180 edited books with Springer (U.K.), Elsevier, Nova Science Publishers, Inc. (U.S.A.), CRC Press Taylor & Francis Asia Pacific, Trans Tech Publications Ltd. (Switzerland), IntechOpen Limited (U.K.), Wiley-Scrivener, (U.S.A.) and Materials Research Forum LLC (U.S.A). He is a member of various journals' editorial boards. He has served as Associate Editor for journals (Environmental Chemistry Letter, Applied Water Science and Euro-Mediterranean Journal for Environmental Integration, Springer-Nature), Frontiers Section Editor (Current Analytical Chemistry, Bentham Science Publishers), Editorial Board Member (Scientific Reports-Nature) and Review Editor (Frontiers in Chemistry, Frontiers, U.K.) He has also guest-edited various special thematic issues for the journals of Elsevier, Bentham Science Publishers, and John Wiley & Sons, Inc. He has attended as well as chaired sessions at various international and national conferences. He has worked as a Postdoctoral Fellow, leading a research team at the Creative Research Initiative Center for Bio-Artificial Muscle, Hanyang University, South Korea, in the field of renewable energy, especially biofuel cells. He has also worked as a Postdoctoral Fellow at the Center of Research Excellence in Renewable Energy, King Fahd University of Petroleum and Minerals, Saudi Arabia, in the field of polymer electrolyte membrane fuel cells and computational fluid dynamics of polymer electrolyte membrane fuel cells. He is a life member of the Journal of the Indian Chemical Society. His research interest includes ion exchange materials, a sensor for heavy metal ions, biofuel cells, supercapacitors and bending actuators.

Ms. Maha Khan is a Research Scholar at the Department of Applied Chemistry, Aligarh Muslim University (A.M.U.), Aligarh, India. She has also pursued her Bachelor's in Chemistry and Master's in Polymer Science and Technology from A.M.U., Aligarh. Her research work focuses primarily on enzymatic biofuel cells, a pathway to clean and green energy.

Dr. Mohammad A. Jafar Mazumder has been serving as a Professor of Chemistry at King Fahd University of Petroleum & Minerals (KFUPM), Saudi Arabia. He has extensive experience in designing, synthesizing, and characterizing various organic compounds, ionic and thermo-responsive polymers for corrosion, water treatment, and biomedical applications. Dr. Jafar Mazumder obtained his B.Sc (Hons.), M.Sc (Chemistry) from Aligarh Muslim University, India, MS (Chemistry) from KFUPM, Saudi Arabia, and Ph.D. in Chemistry (2009) from McMaster University, Canada.

 In more than 20 years of academic research, Dr. Jafar Mazumder has had the opportunity to work with several international collaborative research groups and has exposed himself to a broad range of research areas. Dr. Jafar Mazumder secured 8 US patents, published more than 85 articles in peer-reviewed journals, 37 conference proceedings, 9 book chapters, and co-edited 4 books with Springers and Trans Tech publications. He is awarded as a Fellow of the Royal Society of Chemistry and Chartered Chemist, Association of Chemical Profession of Ontario, Canada. Besides, Dr. Jafar Mazumder is a member of the American Chemical Society (ACS), Canadian Society for Chemistry (CSC), Canadian Biomaterial Society (CBS), and a life member of the Bangladesh Chemical Society (BCS). In his academic career, he was awarded numerous national and international scholarships and awards including the prestigious Indian Council for Cultural Relations (ICCR) Scholarship from Govt. of India for undergraduate studies in India, Aligarh Muslim University undergraduate & graduate Gold medal, and certificate of excellence from the Ministry of Human Resource Development, Govt. of India, and MITACS postdoctoral fellowship (Canada) for pursuing postdoctoral research in Chemical and Biomedical Engineering.

Currently, Dr. Jafar Mazumder is actively involved in several ongoing university (KFUPM), government (KACST, NSTIP), and client (Saudi Aramco) funded projects in the capacity of principal and co-investigators. His current research interest includes the design, synthesis, and characterization of various modified monomers and polymers for

c

potential use in the inhibition of mild steel corrosion in oil and gas industries and the preparation of multilayered polyelectrolyte coated membranes for the removal of heavy metals and organic contaminants from aqueous water samples. The long-term scientific goal of Dr. Jafar Mazumder is not merely to make science fun and entertaining for people. It is to engage them with a multidisciplinary scientific mission at a deeper level to create a space through which they can interact with scientific ideas, develop connections between science, engineering, and biology, and thoughts of their own to contribute to society. He feels this goal and engaging personality make him a pleasant person to work with and help inspire his co-workers in any professional setting.